In Situ Assessment of Structural Timber

RILEM STATE OF THE ART REPORTS
Volume 7

For other titles published in this series, go to
www.springer.com/series/8780

Bohumil Kasal • Thomas Tannert

Editors

In Situ Assessment of Structural Timber

State of the Art Report of the RILEM Technical Committee 215-AST

Springer

Editors

Bohumil Kasal
Fraunhofer Institute for Wood Research
Wilhelm-Klauditz-Institut WKI
Bienroder Weg 54E
38108 Braunschweig
Germany
E-mail: bo.kasal@wki.fraunhofer.de

Thomas Tannert
Berne University of Applied Sciences
Timber and Composite Construction
Solothurnstrasse 102
2500 Biel
Switzerland
E-mail: thomas.tannert@bfh.ch

ISBN 978-94-007-0559-3 e-ISBN 978-94-007-0560-9
DOI 10.1007/978-94-007-0560-9
Springer Dordrecht Heidelberg London New York

Cover design: eStudio Calamar S.L.

Printed on acid-free paper

Springer is part of Springer Science+Business Media (www.springer.com)

Contents

State-of-the-Art Report

This report is partly based on the MS thesis of Gretchen Lear entitled "Improving the Assessment of In Situ Timber Members with the Use of Nondestructive and Semi-Destructive Testing Techniques" (under the direction of Bohumil Kasal) submitted to the Department of Civil Engineering of North Carolina State University, Raleigh, NC, in December 2005.

Contributing TC AST-215 members are: Ronald Anthony, Wood Scientist, Fort Collins, USA; Professor Clara Bertolini Cestari, Politecnico di Torino, Italy; Dr. Milos Drdacky, Institute of Theoretical and Applied Mechanics, Prague, Czech Republic; Professor Bohumil Kasal, Penn State University, USA; Dr. Nicola Macchioni, Istituto per la Valorizzazione del Legno e delle Specie Arboree, Italy; Professor Maurizio Piazza, Università degli Studi di Trento, Italy; Dr. Olivia Pignatelli, Dendrodata s.a.s, Italy; Dr. Mariapaola Riggio, Università degli Studi di Trento, Italy; Dr. Mehdi Sbartaï, Université Bordeaux, France; Dr. Thomas Tannert, Bern University of Applied Sciences, Switzerland; Dr. Nobuyoshi Yamaguchi, Building Research Institute, Tsukuba, Japan.

Dr. Thomas Tannert provided vital help during the final phases of the review. Help of Ronald Anthony with the manuscripts is greatly appreciated.

RILEM TC 215-AST
Bohumil Kasal and Thomas Tannert
Editors

Preface

Wood is one of the most intriguing structural materials and the only one that is truly renewable. Along with stone, wood is the oldest structural material on the planet and has been extensively used throughout the human history. Due to its aesthetical value and positive environmental impact, wood has experienced renaissance in construction. As a biodegradable, hygroscopic, non-isotropic material, the wood presents special challenges for a professional and requires through knowledge ranging from biology to continuum mechanics.

This state-of-the-art report reflects the current knowledge in the area of in situ assessment of the physical and mechanical properties of wood structures. Nondestructive, semi-destructive and destructive methods are described in a systematic manner where technology, equipment and limitations are discussed. Some of the discussed methods are used in other materials such as masonry and concrete. Most of the methods, however, are specific to wood and special qualifications are required to understand and apply these methods effectively. Existing methods are constantly improved and new methods are being developed. This report includes methods that are used in practice or have shown significant promise and have a body of knowledge that supports statements made in this report.

To continue to improve the assessment of in situ timber members, more research is required to both estimate individual member strengths as well as obtain accurate quantification of deterioration. After extensive research of stress wave investigative techniques for timber members, the conclusion was drawn that the relationship between stress wave parameters and timber mechanical properties was not developed enough to accurately predict in situ member strength. Published research and associated correlation values were inconsistent and at times conflicting depending on technique, species and stress wave parameters employed. Research opportunities lie in finding a method of applying stress wave techniques to in situ members and arriving at reliable estimates of the member's strength.

Improvement of the tensile technique could also be addressed. Correcting equipment details as well as addressing sample size for grain inclusion and improvements to the methodology for extracting samples along the grain could result in a more viable technique for establishing tensile strength. Bending strength estimates could then be improved using tensile strength as an estimator instead of core compressive strength.

Accurate quantification of deterioration also needs to be addressed with future research. X-ray and resistance drilling techniques have proven that they can accurately detect areas of deterioration; however these techniques can be improved by adding the ability to quantify and identify different stages of deterioration. Possibilities for X-ray investigation include the ability to assign different density levels to stages of decay as well as being able to accurately measure areas of reduced density, including those shadowed by sound material. Quantitative analysis using digital imaging appears to be a promising technique. Low mass density of wood permits easy application of portable, low energy X-rays. Resistance drilling could be improved by addressing the issue of drilling path deviation associated with the flexible needle so that more accurate plots of member cross-sections can be constructed.

The authors believe that this report will be a useable tool for professionals, researches, educators and students.

Bohumil Kasal

List of Symbols and Abbreviations

ALSC	American Lumber Standard Committee, Inc.
NDS	National Design Specification for Wood Construction
ASTM	American Society for Testing and Materials
λ	wavelength (length)
f	frequency (1/time)
V	stress wave velocity (length/time)
L	length/distance between two points (length)
T	time (time)
δ	logarithmic decrement (amplitude change/cycle)
A_0	initial amplitude measurement (displacement)
A_j	amplitude j cycles apart (displacement)
j	number of cycles between measured amplitudes
α	attenuation factor (amplitude change/unit length)
x	length of stress wave propagation (length)
A_x	amplitude at the propagation length x (displacement)
E_d	dynamic modulus of elasticity (force/area)
ρ	mass density of the member (mass/unit volume)
υ	Poison's ratio
h	Plank's constant
c	velocity of light (length/time)
E	radiation energy (work)
I_X	emergent intensity of radiation beam (rate of energy emission)
I_O	initial intensity of radiation beam (rate of energy emission)
t	thickness of the material (length)
μ	linear absorption coefficient (1/length)
Ug	geometric unsharpness (length)
S	size of the focal spot within an X-ray tube (length)
a	distance from the source to object (length)
b	distance from the object to imaging material (length)
RM	Resistance Measure (length^2)

h	height of specimen (length)
LVDT	Linear Variable Differential Transducer
f_c	compressive strength of the core (force/area)
$F_{\max}$	failure load (force)
l	length of core sample (length)
d_c	diameter of the core (length)
r^2	correlation coefficient
E_s	static modulus of elasticity (force/area)
E_d	dynamic modulus of elasticity (force/area)
F_{bending}	bending s trength (force/area)
$F_{\text{compression}}$	compressive strength (force/area)
F_{tension}	tensile strength (force/area)
WCLB	West Coast Lumber Inspection Bureau
$x_{0.05}$	ASTM lower 5th percentile strength calculated using the t-distribution (force/area)
$\overline{x}_{\text{ASTM}}$	average ASTM compression specimen response (force/area)
$t_{\alpha,n-1}$	critical value associated with a given probability level, α, and degrees of freedom v
SD_{ASTM}	standard deviation of ASTM compression specimens (force/area)
$\overline{x}_{\text{core}}$	average response of cores (force/area)
SD_{core}	standard deviation of core response (force/area)
f_t	tensile strength (force/area)
b	base dimension of the tensile specimen (length)
h	height dimension of the tensile specimen (length)
$\overline{x}_{\text{tension}}$	average micro-tension specimen response (force/area)
$\text{SD}_{\text{tension}}$	standard deviation of micro-tension specimens (force/area)

List of Figures

List of Tables

Introduction

The need for structural assessment can arise from multiple motivations such as performance reports to address structural adequacy, historic preservation and building change of use. The time and cost of inspections is justified with the assurance of safety gained, the protection of capital investments and minimization of costs involved with maintaining the structure.

For historic structures, quality assessments of members allow for the maximum retention of original material. The preservation of original structural fabric and associated construction conserves both the cultural significance of the building including architectural qualities and building techniques and the historic and socially important aspects associated with the structure. Furthermore, gaining additional understanding of building material durability, capacity, behavior and use, as well as building techniques and craftsmanship from existing structures provides knowledge that can be applied to present-day construction.

A quality assessment begins with the assessment of the members and components that make up the structure as a whole. A visual inspection of the member under consideration is performed and naturally occurring characteristics and deterioration are inventoried. Natural characteristics include features such as knots, slope of grain and seasoning checks while deterioration includes damage from insect infestation or fungal decay and potential damage due to progressive failure. After the visual inspection, in situ grading provides a structural grade based on the size, number and location of growth characteristics according to the member's size and structural use. The structural grade is assigned to a section without the portion decayed by biotic agents.

A common practice in evaluating and assessing timber members in existing structures, in situ grading allows for the assignment of structural grades through the application of grading rules. Grading rules are published by a variety of organizations and written for different species and regions, such as [1, 2]. Grading organizations that write grading rules and the grading rules themselves are certified and monitored by different national standardization bodies that are typically coordinated at an international level, such as the CEN in Europe.

B. Kasal and T. Tannert (eds.), In Situ Assessment of Structural Timber, 1–3.

After establishing the member grade with appropriate grading rules, design values for visually graded timber published at national level, e.g. National Design Specification (NDS) for Wood Construction [3] are referenced to establish mechanical property design values. All of the reference design values are categorized by species, size and grade. Most of the visually graded dimension lumber design values are based on the testing of full size specimens in accordance with various national standards such as ASTM D1990 [4], while visually graded timbers, decking, and some species and grades of dimension lumber are based on the provisions of ASTM D245 [5].

ASTM D245 prescribes for the adjustment of clear wood property values, provided by test methods in ASTM D2555 [6], by strength ratios and adjustment factors. Strength ratios and adjustment factors are based upon natural characteristics present within the material, environmental conditions and the intended structural use. This practice of reducing clear wood values is based on extensive research including tests of small clear specimens and of full-sized members as well as detailed studies of strength and variability of clear wood and the effect of various factors and defects on those properties [5].

There are several drawbacks associated with this type of assessment and assignment of design values to in situ timbers. The published NDS design values are based on the testing of new timbers and may not represent mechanical properties typically found in older timbers. At this point, new timber that is harvested for construction is second generation growth. It is fast grown to produce wood quickly, and the result is reduced density and mechanical properties. In older existing structures it is common to find first generation growth. These timbers were harvested from forests that grew at a much slower rate, and most likely have different density and mechanical property values. The application of current NDS design values does not consider this factor and can underestimate the member's mechanical properties.

In addition, allowable stress values published in NDS and based on ASTM D245 practices are based on the clear wood values that are representative of an entire species or species group and not individual members. In doing so, the values must account for the natural variability within the species and conservative property values are published to account for weak members. Excluding modulus of elasticity and compression perpendicular-to-grain values, the resulting values from ASTM D245 are based on a 5% exclusion limit; meaning that 95% of members in a species group and stress grade are expected to have strengths that are above that established by ASTM D245 and published in the NDS. Assessment deals with individual members (or small groups) and not whole species groups, so the application of design values based on the species as a whole may be inappropriate.

Not only do current inspection techniques and in situ grading run the risk of underestimating individual member properties, member capacities can also be overestimated if areas of deterioration are not located and the extent of the deterioration quantified. Detection and quantification of biotic deterioration is difficult or even impossible through visual inspection. Biotic deterioration refers to the damage caused by the attack of living agents such as bacteria, fungi, or insects. Loss in weight, stiffness and strength can occur before decay can be visually detected. Mechanical properties can be reduced by 10% before any visual indicators are present, and when a weight loss due to decay of only 5–10% is detected, the loss in mechanical properties can be as large as 80% [7].

The assessment of in situ timber members can be improved in two ways: by gaining more accurate estimates of individual member strengths and by locating and quantifying deterioration. For this study, several techniques were researched and applied to members to predict individual member strengths or locate and quantify deterioration. This study includes a description of the nondestructive and semi-destructive testing methods investigated, details of testing, results and a discussion of the findings.

Non-destructive and semi-destructive testing techniques were researched for their ability to improve the assessment of in situ timbers. Background information on their technology and application, with respect to strength assessment and/or deterioration detection and quantification, is discussed for each.

References

1. Standard No. 17: Grading Rules for West Coast Lumber. West Coast Lumber Inspection Bureau, Portland, 1996.
2. Standard Grading Rules for Southern Pine Lumber. Southern Pine Inspection Bureau, Pensacola, 2002.
3. Supplement: National Design Specification. Design Values for Wood Construction. American Forest & Paper Association, Inc., 2005.
4. ASTM D1990-00. Standard Practice for Establishing Allowable Properties for Visually Graded Dimension Lumber form In-Grade Tests of Full-Size Specimens. ASTM International.
5. ASTM D245-00, Standard Practice for Establishing Structural Grades and Related Allowable Properties for Visually Graded Lumber. ASTM International.
6. ASTM D2555-05a. Standard Practice for Establishing Clear Wood Strength Values. ASTM International.
7. Stalnaker, J.J. and Harris, E.C. *Structural Design in Wood*, 2nd ed. Kluwer Academic Publishers, Boston, 2002.

Chapter 1
Stress Waves

Bohumil Kasal, Gretchen Lear, and Thomas Tannert

Abstract This chapter describes principals of the stress wave applications to timber structures including sonic and ultrasonic techniques. Correlations between measured physical parameters and material properties are discussed. Stress waves can be applied in transverse and longitudinal directions and parameters such as velocity; time of flight and frequencies can be analyzed and correlated with elasticity and strength properties of investigated structural members. While correlation between dynamic and static modulus of elasticity is relatively good, correlations between the dynamic modulus of elasticity and other mechanical properties are poor. The methods require knowledge of material density and moisture contents.

Key words: sonic waves, ultrasonic waves, ultrasonic echo, ultrasonic tomography, stress waves, dynamic modulus, mechanical properties, density, velocity, frequency, spectrum, attenuation, correlation

1.1 Background

Stress wave investigation has long been used for non-destructive investigation of wood members. The time of flight or the corresponding calculated velocity is the most common stress wave parameter used during timber investigations, however, attenuation and frequency spectrum analysis can be used as well.

The use of stress waves in non-destructive testing is based on the propagation of sound waves through material and is widely used for detecting interior voids and deterioration in structural members, as well as for mechanical property measurement. Sonic stress waves, commonly referred to simply as stress waves, are those with frequencies within the audible range. Ultrasonic stress waves are inaudible, having frequencies above 20,000 Hz.

Waves are transmitted by elastic materials, and propagate through a material by means of the oscillatory motion of the material particles. Wave forms are defined

B. Kasal and T. Tannert (eds.), In Situ Assessment of Structural Timber, 5–24.

by the particle motion relative to the wave propagation through the material. The most commonly used wave form is the longitudinal, or compression, wave in which particles oscillate in the same direction as the wave propagation. Transverse waves cause particle oscillation perpendicular to the direction the stress wave is moving.

Some of the characteristics of sound waves are frequency, the number of oscillations per given time increment, and the wavelength, which is the distance taken for the completion of one cycle. Wavelength is inversely proportional to the frequency; therefore low frequencies correspond to longer wavelengths and high frequencies with short wavelengths. Wavelength λ is related to the frequency, f, and velocity of the wave, V, as follows:

$$\lambda = \frac{V}{f} \tag{1.1}$$

Speed and attenuation of sound waves are the primary parameters used for non-destructive evaluation. Wave speed is simply determined through the relationship:

$$V = \frac{L}{T} \tag{1.2}$$

where V is velocity, L is the distance between two points along the path of the sound wave, and T is the time taken to traverse the distance L. The velocity of sound in the material can then be used to predict mechanical properties through empirical relationships or indicate voids and/or deterioration and will be discussed further in coming sections.

Attenuation, which is the amplitude loss of the stress wave, results from two sources; scattering at material interfaces and absorption. Attenuation, or damping, of the stress wave as it propagates has been shown to have correlation to material strength [1]. Kaiserlik [2] presented data from experiments where the average rate of attenuation was used in a model with other parameters to predict tensile strength. The model improved the estimation of tensile strength over traditional empirical relationships from a correlation coefficient of $R^2 = 0.697$ to $R^2 = 0.819$.

The rate of attenuation can be express by logarithmic decrement, the loss of amplitude per sinusoidal cycle, and is given by Equation (1.3) where δ is the rate of decay, A_0 and A_j are the amplitudes of two waves j cycles apart.

$$\delta = \frac{1}{j} \ln \frac{A_0}{A_j} \tag{1.3}$$

Attenuation can also be characterized with a similar but slightly different parameter, attenuation factor. The attenuation factor is the loss in amplitude per unit length of propagation for an advancing stress wave. Equation (1.4) defines the attenuation factor, α, where x is the length of propagation, A_0 is the initial amplitude, and A_x is the amplitude at point x.

$$\alpha = \frac{1}{x} \ln \frac{A_0}{A_x} \tag{1.4}$$

1.2 Equipment

1.2.1 Sounding

Sounding is one of the oldest methods used to inspect in situ timber members and provides a quick inspection procedure to identify serious decay within members. With sounding, the timber member is struck by a blunt object, typically a hammer. From the resulting sound tone, a trained inspector can make inferences to the member's condition. This method has the advantage of being able to rapidly screen timber members, however, it is highly subjective and diagnosis can vary between inspectors. Sound quality can be affected by factors other than decay which can complicate the interpretation. In addition, sound interpretations cannot quantify the extent of decay, and sounding only indicates serious decay, not initial or moderate, even when experienced persons perform the tests. In order to address these drawbacks of the sounding methods, more technical experiments using stress waves can be conducted.

1.2.2 Sonic Stress Waves

1.2.2.1 Velocity Measurement

Sonic stress waves can be imparted into a member with a simple mechanical impact from a hammer or blunt object in the transverse or longitudinal direction. Two accelerometers, mounted in the impact device and/or placed on the member, are used to detect passing stress waves and record time measurements.

Transverse wave investigation requires access to two opposing faces of the member, as illustrated in Figure 1.1. The impact device contains an internal accelerometer. At the point of impact when the stress wave is induced, the accelerometer starts the timer. A second accelerometer is situated on the opposite face of the member and stops the timer when the wave front is detected. The thickness of the member is equal to the path length the stress wave travels and is used as the length parameter to calculated wave velocity, see Equation (1.2).

Longitudinal stress waves can be imparted in two ways, by a direct impact at the end face or impacting transducers embedded in the member, see Figures 1.2 and 1.3. As the stress wave passes the first accelerometer a timer is started, then upon reaching the second accelerometer the timer is stopped. The average velocity can be calculated with the recorded time and measured distance between the accelerometers. When using embedded transducers, the angle between the transducer and member can affect the transit time if it is too large and should be considered when testing. More information on this technique can be found in [3].

Hammers are the typical devices used to impart mechanical stress waves in members. The material of the hammer head will change the frequency of the wave which

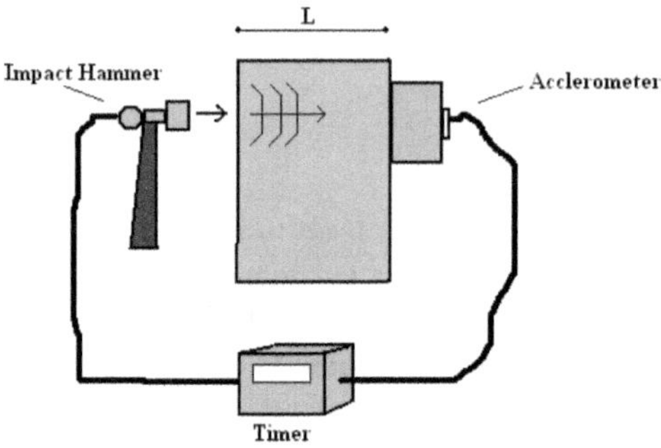

Fig. 1.1 Configuration for stress wave testing using impact hammer in transverse direction

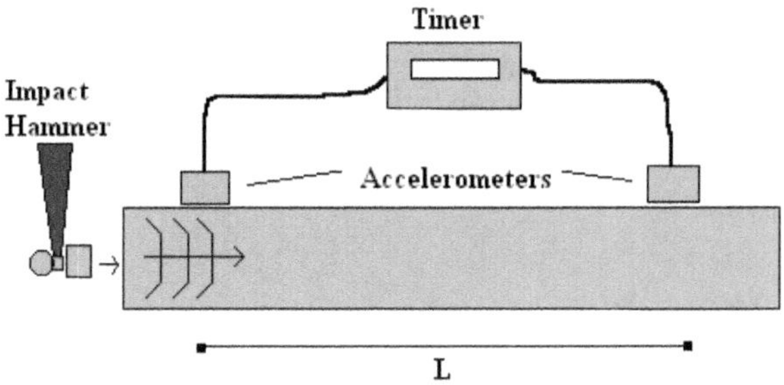

Fig. 1.2 Configuration of stress wave testing using impact hammer in longitudinal direction, end face impact

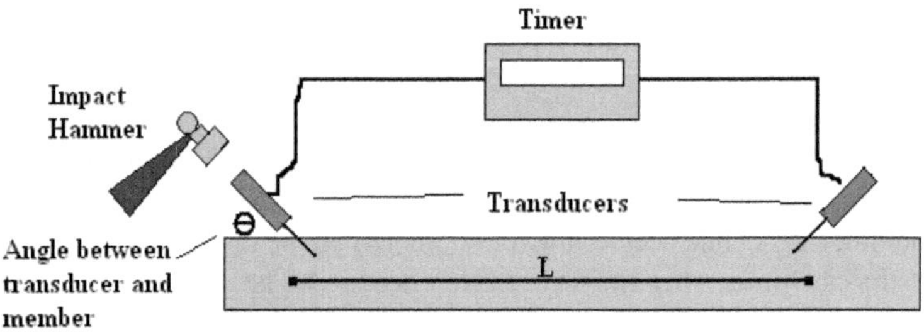

Fig. 1.3 Configuration of stress wave testing using impact hammer in longitudinal direction, transducer impact

it produces; softer materials will produce lower frequency stress waves, and conversely harder materials will induce higher frequencies. Additionally, the weight of the hammer will affect the frequency; heavy hammers will produce lower frequencies than light hammers because they experience a longer time of contact with the surface on impact [4].

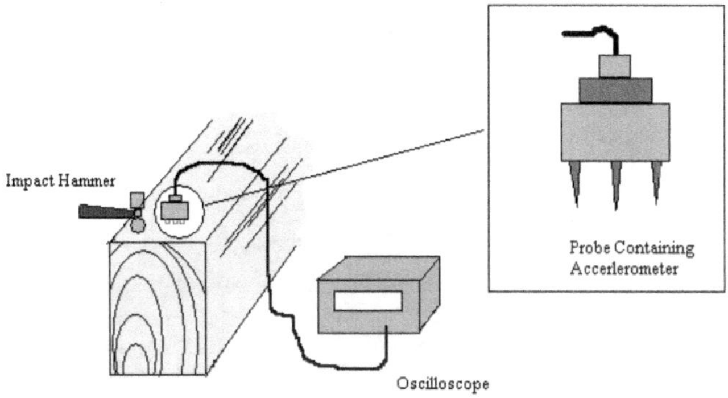

Fig. 1.4 In situ stress wave set up for frequency spectrum data [5]

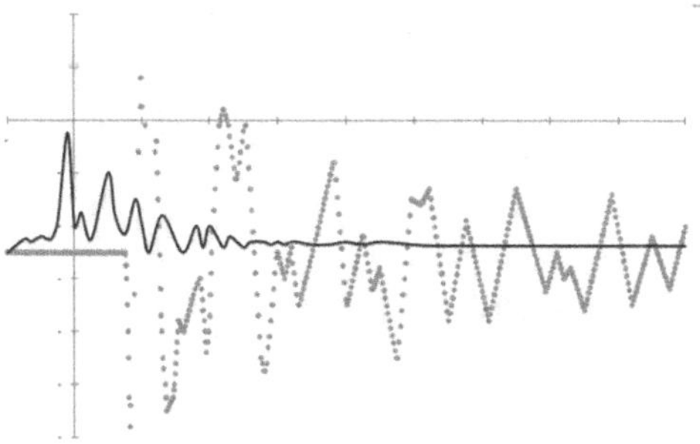

Fig. 1.5 Sample time-domain wave form converted into the frequency spectrum [5]

1.2.2.2 Frequency Spectrum Analysis

Frequency spectrum analysis can be used to assess and quantify timber decay and has the advantage over traditional velocity measurement techniques in that only one member face is required for testing. As described in [5] and illustrated in Figure 1.4 a specially designed probe containing an accelerometer and wired to an oscilloscope is attached to the member face. A hammer is used to induce the stress wave into the member. As the probe receives the stress wave signal, the oscilloscope transforms the time-domain signal into a frequency spectrum through Fast Fourier Transformation (Figure 1.5). Frequency is dependent on timber condition, therefore frequency ranges can be designated to levels of condition and provide an objective method for characterizing and differentiating degrees of deterioration. Multiple impact records can be compiled and plotted to produce three-dimensional plots of member cross-sections, illustrated in Figure 1.6 [5].

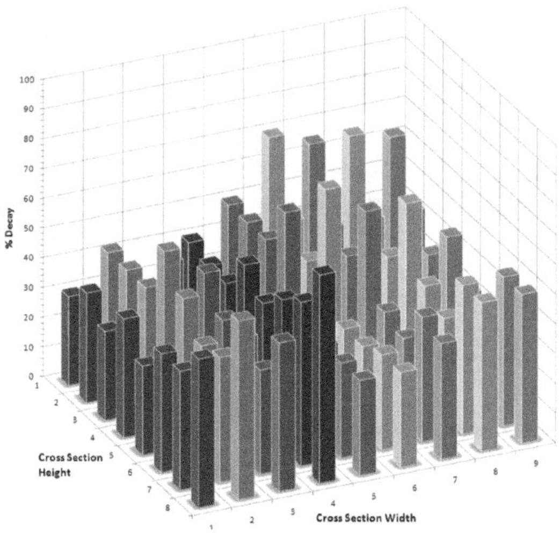

Fig. 1.6 Frequency spectrum analysis: three-dimensional plot of the percentage of member cross section subjected to decay [5]

1.2.3 Ultrasonic Technique

The most common form of ultrasound utilizes a piezoelectric material, usually a quartz-crystal, which converts a supplied electrical current into oscillatory waves. During operation the transducer is supplied with an electrical current, converts it into a wave signal, and directs the wave into the material. A data acquisition system consisting of a signal detector, an analyzer, and a recorder collects and retains the wave information for further analysis [6].

Contact scanning requires transducers to be coupled to the material surface for direct transmission measurements. Air transmits sound waves poorly, so couplants are used to create complete contact between the transducers and material surface to limit signal retardation. The amount of energy that is transmitted into the material is proportional to the coupling force applied to the transducer [7]. Common couplants include water, oils, greases, pastes and sometimes rubber when the transducers are held in place by hand. In some cases, the surface of the member may need preparation to ensure adequate coupling of the transducers. This can include planing, sanding or other methods of smoothing the surface.

Non-contact scanning uses transducers that do not require contact with the material surface to transmit ultrasonic waves into the material. Air can be used as the couplants in certain cases, which has the advantage of avoiding damage to the surface. Because of the difference between air and solid material impedance, the technology can be sensitive to surface roughness and experience significant energy reflection [8].

Testing frequencies are typically in the range of 20 to 500 kHz and are kept relatively low because of the high wave attenuation in timber due to its heterogen-

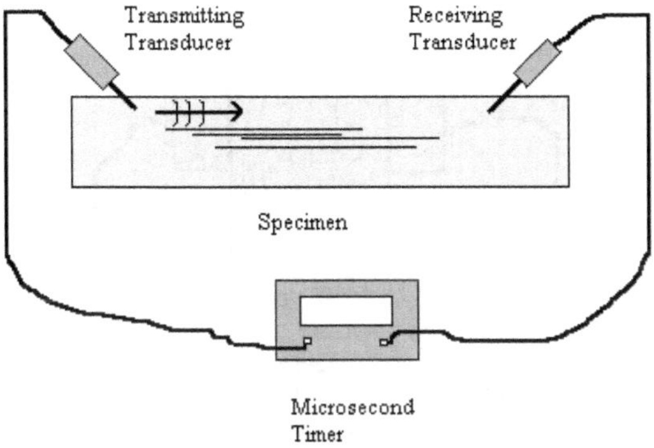

Fig. 1.7 In situ ultrasonic testing in the longitudinal direction, end access is not required

eous nature. Ultrasonic investigation is commonly used in timber grading with the transducers arranged at the ends of the lumber pieces to measure wave propagation directly along the fibers. For in-situ evaluation, that transducer configuration is not advantageous considering access to the member ends is most often limited. To address this limitation the transducers can be mounted on the same side of the member as shown in Figure 1.7. A signal is induced along the length of the member, transit time and pathlength are recorded, and velocity calculations are made. This configuration will neglect the condition of the member ends; however, it provides longitudinal ultrasonic stress waves and parameters to estimate global properties of the member.

Transducers can also be arranged to impart a signal directly through the member as shown in Figure 1.8. In this case, two transducers are oriented across from each other on opposite surfaces of a member, a signal is induced directly through the member, transit time and pathlength are recorded, and velocity calculations can be made. Access to opposite sides of an object is required for this type of configuration which can limit its use for in situ investigations.

1.2.4 Ultrasonic Echo Technique

Ultrasonic echo technique can be applied on wood with longitudinal and transverse waves. The technique is based on the reflection of the acoustic waves on material inhomogenities like the back surface of the specimen or other interfaces. With this technique it is possible to get indirect information about the condition and internal damage of the material. The specific characteristics of wood require signals with low-frequency (50–200 kHz). Recent research [9–13] applied the method and showed that low frequency shear waves are suitable to assess wood.

Figure 1.9 shows a schematic representation of the basic principle. The set-up consists of a signal generator, a preamplifier, an amplifier, a transducer and a PC for

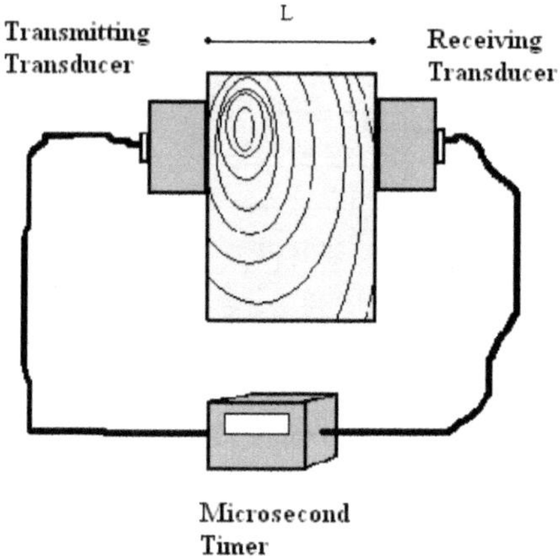

Fig. 1.8 Ultrasonic transverse application

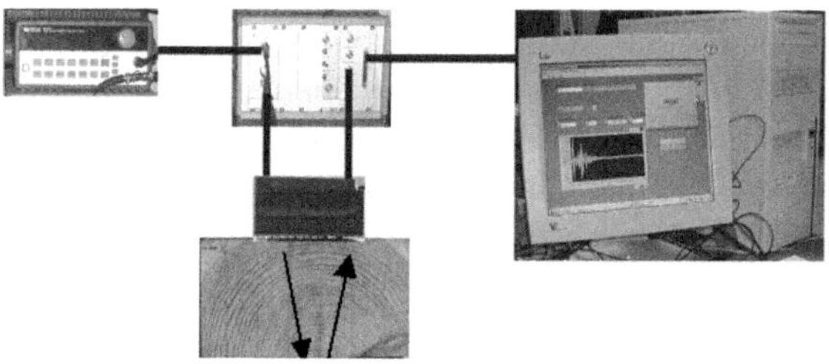

Fig. 1.9 Setup for using ultrasonic echo technique [10]

the data equalisation. During the measurement, the sensor is placed on a surface of the element. The sonic wave, illustrated by the arrows, passes through the element and is reflected at the back wall. The reflected signal is called back-wall echo and is received by the sensors. Any structural irregularity in the wooden structure produces a change in signal structure of the back wall echo.

A sensor head developed by ACSYS [14], has 24 sensors; 12 sensors act as transmitters and 12 sensors act as receivers, see Figure 1.10. The simultaneous activation of all transmitting sensors reduces coupling problems on rough wooden surfaces and reduces the spread of the sound field when compared to a single sensor [10]. Another advantage is that no coupling agent is necessary and thus the surface does not get polluted. The probes are activated with a frequency of 55 kHz, which for pine and measurements perpendicularly to the fiber results in a wavelength of 25 mm (1 in). Longitudinal wave transducers can also be applied, but these have to be used with a coupling agent like Vaseline or glycerine.

Results using the ultrasonic echo technique are often presented as A- and B-scans. The A-scan (Figure 1.11, right) shows the transmission time and intensity of the pulse; while the B-scan (Figure 1.11, left) is a composition of various A-scans that are recorded with a defined distance. The B-scan is a 2D cross section through

Fig. 1.10 Sensor head for ultrasonic echo technique [11]

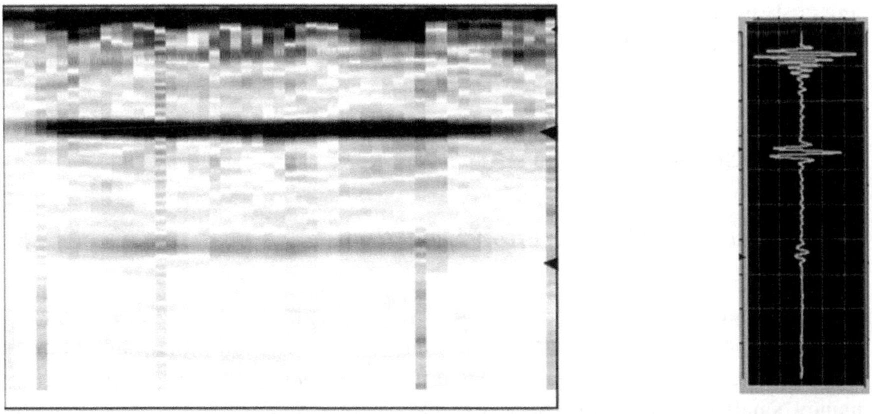

Fig. 1.11 Result of a measurement along specimen, left as a B-scan, right as an A-scan [10]

the specimen and enables to identify a change in signal structure along the measured axis. If several B-scans that are recorded with a defined offset are combined, it is possible to interpolate a horizontal layer. These interpolated layers (C-scan) can give 3D information about the structure and damages in the object [12].

The ultrasonic echo technique enables to detect the back walls as well as irregularities inside wooden structures. Relative results can be calibrated by measuring the dimension of the specimen and by using other assessment methods on areas were this is possible. When calibrated values are available, it is possible to determine the thickness of beams where the back wall is not accessible and to determine the dimensions and inner structure of structural elements. Detailed tests are only necessary in areas with no or an unexpectedly early echo from the back wall, because this can be a sign of damage. An automatic scanning system can be arranged for structural elements with slender geometry [12].

The ultrasonic echo technique allows the direct localization of a reflector, like a back wall or any inhomogeneity like damage in the wooden structure. It is difficult, however, to locate the exact position of damage within the specimen. It is also difficult to distinguish between one large irregularity like a knot or a cluster of small ones. Nevertheless, it can be assumed that a clear back wall echo shows that the specimen is free of defects [10].

The resolution of the method strongly depends on the wavelength in the material. As an approximation, it is not possible to detect damages with a size less than a half the wavelength. Since the wave length in wood is 25 mm (1 in), it follows that it is not possible to use direct echos to detect damages with dimension smaller than 12 mm (0.5 in). An increase of frequency is limited because higher frequencies result in higher attenuation of the ultrasonic signal with increasing current time [10].

Systematic measurements on specimens with artificial imperfections are required to make qualitative conclusions from measuring results. An exact classification of damage by analyzing the signal structure is not yet possible. The potential in reducing costs by using the method to determine damages prior to a renovation process needs to be evaluated [12].

1.3 Application

1.3.1 Defect and Deterioration Detection

For the detection of voids and defects, wavelengths play a key role. In general, defects that are smaller than half the wavelength of the induced signal cannot be detected by stress wave investigation. Therefore, ultrasonic stress waves (higher frequency, smaller wavelengths) have a greater ability to detect subtle interior voids and defects in materials than sonic stress waves. However, ultrasonic stress waves have a higher susceptibility to wave attenuation and signal loss which can limit their use in detecting and quantifying deterioration and voids. Sonic stress waves have longer wavelengths and are not as sensitive to smaller defects and are subsequently more useful in identifying larger, more significant voids and deterioration.

1.3.1.1 Velocity Measurement

The velocity of wave transmission can give an indication of material condition. As materials deteriorate, their stiffness is reduced. Wave velocity is proportionally linked to the square root of the material stiffness in which it is induced. Slower velocities or a longer transient time when compared to sound material indicate possible deterioration. With multiple scans, transient times can be mapped on the member, or 2D and 3D plots can be constructed of the member surface to assess the condition. Longer transient times would be seen as high points on plots and would indicate deterioration or voids; see Figure 1.12 for an example plot of stress wave times taken along the length of a member (not actual data). It is critical to calibrate the stress

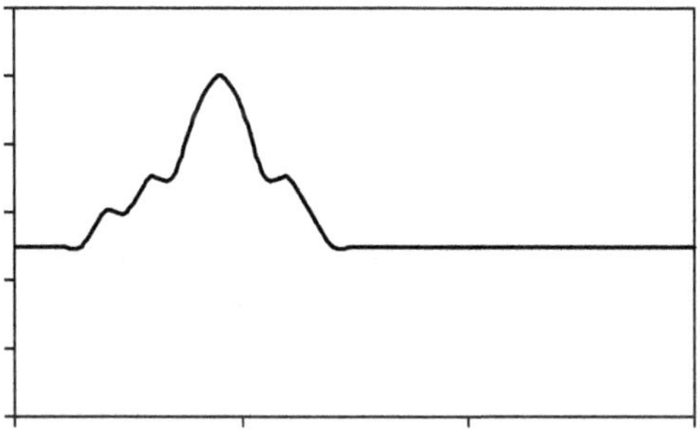

Fig. 1.12 Example plot of stress wave testing

wave timing mechanisms to sound material in order to establish a basis of comparison and velocity measurements must be done in a consistent manner to minimize the variability in testing and results [15].

Stress wave velocities in timber members are directionally dependent. Wood structure is mostly made up of long, tubular cells oriented longitudinally. For softwoods approximately 90% and for hardwoods 80–95% of the cells are oriented in the longitudinal direction. Rays, which are oriented in the horizontal axis [16], make up the remaining percentage.

In sound wood, longitudinal transmission velocities generally fall within the range of 3500–5000 m/s while in the transverse direction velocities are 1000–1500 m/s [3]. Stress waves induced in the longitudinal direction have a higher velocity because they travel along the vertically oriented cells and encounter few to no boundaries to impede their progression. Transverse waves encounter numerous boundaries and interfaces at the cell walls which reduce their velocity.

Longitudinal stress waves travel the span of the member and give the average velocity along the length. This velocity measurement is a global parameter which can be used to estimate properties of the member as a whole. Transverse waves travel over a localized portion of a member; their velocities are local parameters and are used to evaluate only local properties and conditions at the testing site.

The speed of stress wave transmission in degraded timber members is reduced as compared to sound members. A 30% reduction in the transmission time implies approximately 50% loss in strength, while a 50% reduction in the transmission time indicates a severely decayed member and extreme loss in strength [3]. In order to compare transmission speeds, sound timber speed must first be established for a base line to compare degraded member sections. Tables of stress wave velocities (for both the longitudinal and transverse direction) for sound timber of different species have been compiled based on experimentation that can be referenced as a source of comparison; however member specific measurements are superior as variability exists within species.

Transverse paths over the cross section of members are more advantageous for the detection of decay as longitudinal waves can bypass areas of deterioration. Multiple transverse stress wave readings can be set up in a grid system, spanning the

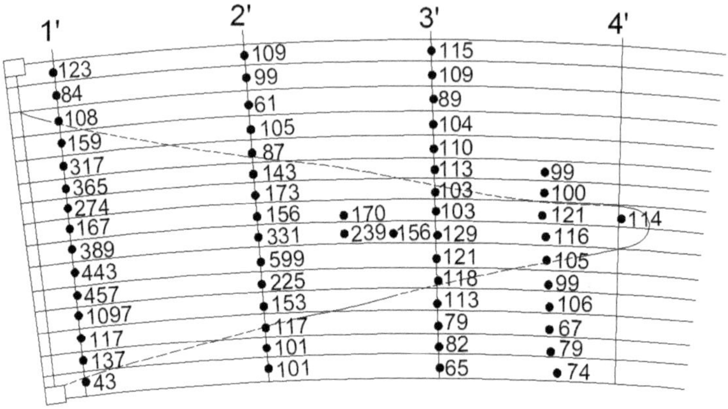

Fig. 1.13 Example mapping of deteriorated regions using stress wave transmission times [17]

horizontal and vertical directions of the member to map out areas with longer transit times and suspected deterioration. Areas of concern can then be analyzed with a finer grid to define the boundary of the deteriorated regions. An example mapping from [17] is shown in Figure 1.13.

1.3.1.2 Attenuation

Deterioration can be identified by the degree of stress wave attenuation. As stress waves travel through a sound member, the wave amplitude reduces with time in a steady manner and the sinusoidal waves are equally spaced. In a degraded member the amplitude of the waves will decrease at a more rapid pace as energy is lost at a higher rate to reflection and absorption [18]. A representation of wave behavior in both a sound and degraded member is shown in Figure 1.14.

Directional propagation considerations, with respect to grain orientation, should be made for timber applications. The rate of attenuation increases when the number of material boundaries encountered is high. Transverse stress waves will meet boundaries at every cell wall and will experience more attenuation than waves traveling parallel to the grain and along the long tubular cells which are only partially interrupted by medullary rays [19].

1.3.1.3 Frequency Spectrum Analysis

Stress wave frequency spectrum analysis can provide extensive information on the deterioration level of a member. As explained in [5], the natural frequency of the wood is determined by the properties of the wood structure, and is inversely proportional to the square root of the mass and directly proportional to the square root of the stiffness. Because decaying wood loses stiffness and mass, the natural frequency

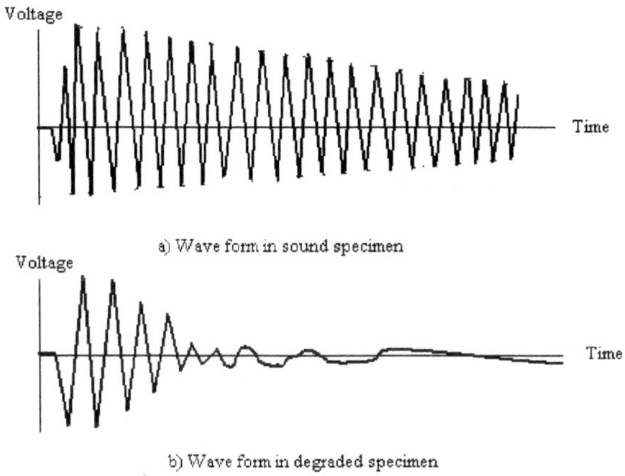

Fig. 1.14 Example stress wave behavior: (a) sound member; (b) degraded member

drops below that of sound wood. Different frequencies are produced based on the amount of deterioration within the member; even incipient decay, which can be difficult to determine with velocity measurements, can be detected. The amplitude of the stress waves can also be a direct indicator of the portion of the wood subjected to different levels of deterioration. As a result of multiple scans, contour maps of the member's cross section can be produced to quantify and plot the extent and location of deterioration.

Plots of the frequency spectrum can be used to detect and quantify the amount and distribution of timber deterioration. The area under the frequency spectrum curve, and within a frequency range associated with a deterioration level, has a positive correlation with the amount of wood subjected to that deterioration level. The combination of the frequency plots and the three-dimensional plots of the cross sections allows the distribution of varying decay levels and sound wood to be quantified and provides a basis for quantitative evaluation of the residual strength of the wood member [5].

1.3.1.4 Ultrasonic Tomography

Cross-sectional images of timber members can be produced with the data collected during ultrasonic investigation. Two-dimensional images are produced by compiling reflection data collected by illuminating the sample with multiple scans from different directions around the member. All characteristic parameters, time of flight, amplitude, frequency, etc., can be used in the data collection to produce the images. One of three algorithms (transform technique, iterative technique or direct inversion) is then applied to the collected data to produce the tomographic images. The ultrasonic tomographs can reveal structural elements in lumber such as knots, grain deviation, cracks, compression wood, fungal attacks and voids [20]. See [8, 20–22] for further information on equipment, application and examples.

1.3.2 Mechanical Properties

No direct relationship between stress wave parameters and material strength exists in current theory, which is one of the fundamental challenges facing implementation of the technique. The estimation of mechanical properties is done through empirical relationships related to stress wave velocities, dynamic modulus of elasticity or stress wave attenuation established through experimental research. This is a practice which is generally accepted as long as strong correlations can be shown [23,24]. The prediction of mechanical properties using stress wave parameters is largely based on correlations between mechanical properties and the modulus of elasticity. Stress wave analysis can effectively measure modulus of elasticity; however it requires significant extrapolation of measurements to predict ultimate strength [23].

1.3.2.1 Velocity Measurement

Sound travels at different speeds through different materials and is affected by the modulus of elasticity and the density. Poisson's ratio also has a minor effect on the wave velocity. After establishing stress wave velocity in a member with the relationship presented in Equation (1.5), the relationship below can be used to calculate the dynamic modulus.

$$E_d = V^2 \rho \times \left[\frac{(1+\upsilon)(1-2\upsilon)}{(1-\upsilon)} \right] \tag{1.5}$$

This equation, where E_d is the dynamic modulus, V is the wave velocity, ρ is the mass density, and υ is Poisson's ratio, is the representation of wave propagation in a 3D medium. The simplified 1D wave form equation for a homogeneous and isotropic material is commonly used for dynamic modulus of elasticity estimations without the consideration of Poisson's ratio. Although wood is neither homogeneous nor isotropic, several researchers have found that 1D theory is appropriate for describing wave behavior in timber [25].

$$E_d = V^2 \rho \tag{1.6}$$

The density of the member may be difficult to determine and may require sampling for laboratory testing. A small sample will only give local information on member density so multiple samples along the length of the member will be required for estimating the global, overall density of the member. After estimates of E_d have been made, empirical relationships are used to estimate timber mechanical properties. These experimentally established empirical relationships have varying correlation values, some of which can be strong and others weak. The complex nature of wave

Table 1.1 Correlations between dynamic and static modulus of elasticity

Material	Static Loading	r^2
clear wood	compression	0.96
clear wood	bending	0.96
lumber	bending	0.92
lumber	bending	0.81–0.85
knotty lumber	bending	0.76
clear lumber	bending	0.90
preservative treated piles	bending	0.58
old lumber		0.65
new lumber		0.70

propagation in timber influences the ability to predict strength properties and the presence of naturally occurring defects influences the ability to predict the performance of timber members [1].

As an alternative to using empirical relationships between stress wave parameters and mechanical properties, sampling can be used to calibrate stress wave measurements with strength properties. Stress wave measurements can be made and followed by sample extraction at the same location. The samples can then be tested destructively to correlate the stress wave parameter to strength. This relationship can then be applied to remaining members under investigation. The correlation between the sample strength and stress wave parameters will vary and is highly dependent on the number of samples taken to establish the correlation. Strong correlations have been shown when adequate sampling to establish the relationship to stress wave parameters is done [26]. Table 1.1 shows the correlations between static and dynamic modulus of elasticity. While the relationships shown in Table 1.1 have some reasonable values of correlation coefficients, it is also indicated that strength estimates are inaccurate.

1.4 Limitations

Several factors can effect the transmission of stress waves in timber, control stress wave application parameters and make interpretation of results difficult. Frequency selection may be subject to the material condition and defect size. Wood has a higher rate of attenuation then most structural material but deterioration will cause increased attenuation effects. Attenuation can be so severe that signals can be completely lost or used only on small member sections. Attenuation will worsen with high frequencies, but high frequency transmissions are also more sensitive to internal defects. The result is that higher, more sensitive transmissions are limited to spans sometimes too short for any practical use. Low frequency transmissions will not experience as much attenuation and can span greater distances; however they are less sensitive to small defects within the member.

Table 1.2 Correlations between static modulus of elasticity and mechanical properties

Property	Static Loading	r^2
Es to F-bending (flatwise)	flatwise	0.52–0.62
Es to F-bending (flatwise)	flatwise	0.56
Es to F-bending (flatwise)	flatwise	0.58
Es to F-bending (flatwise)	flatwise	0.58
Es to F-bending (flatwise)	flatwise	0.77
Es to F-bending (flatwise)	flatwise	0.72–0.74
Es to F-bending (edgewise)	flatwise & on edge	0.42
Es to F-bending (edgewise)	flatwise & on edge	0.32
Es to F-bending (edgewise)	flatwise & on edge	0.46
Es to F-bending (edgewise)	flatwise & on edge	0.41
Es to F-bending (edgewise)	flatwise & on edge	0.64–0.76
Es to F-bending (edgewise)	flatwise & on edge	0.44–0.64
Es to F-bending (edgewise)	flatwise & on edge	0.48–0.71
Es to F-bending (edgewise)	flatwise & on edge	0.45
Es to F-compression		0.71
Es to F-compression		0.61
Es to F-compression		0.45
Es to F-compression		0.62–0.67
Es to F-tension		0.55–0.67

Characteristics of the member such as its geometry, ring orientation and preservative treatments, the physical properties of transducers, the mechanical loading on the member, and its moisture content and temperature [19] can all affect stress wave velocity and attenuation and should be carefully considered. Equipment and measurement conditions should also be taken into account, including the coupling agents and their bond ability, the sensitivity and frequency response of transducers, the difficulty in controlling impact durations and energy from impact hammers, and the frequency selection for parameters investigated. More information on the specific effect of these factors and other influences beyond the scope of this paper can be found in a multitude of publications including [2, 4, 5, 19, 30].

Access to in situ members may also limit the use of some stress wave testing techniques. This would include both the access to appropriate member faces as well as issues related to location, such as areas that would require scaffolding or those that could make proper equipment use difficult. Environmental conditions, such as heat and moisture, and the surface condition of the member can make application of stress wave equipment problematic and can affect transmission data making repeatable experiments difficult.

Errors can arise in ultrasonic testing due to the inadequate coupling of transducers. Air is a poor transmitter of sound waves, so the transducers must have adequate surface contact with the testing member to limit the retardation of the transmission. Common couplants include materials which can adversely affect the surface of the member in question. One must consider the effects of such couplants on the member associated with the surface finishes, chemical reactions that might incur, and any damage that might be the result of removal or clean up. In addition, some surface preparation is commonly called for that can alter the appearance of the member including sanding or leveling of the transducer site to promote ultrasonic transmission.

1.4.1 Deterioration and Decay Detection

Longitudinal waves can give realistic average values of the beam stiffness but can be significantly affected by defects such as ring shakes, localized decay and moisture gradients. The waves can bypass these areas, recording only the fastest transit times [31]. Stress wave times are also unaffected by damage caused by termites as shown in [32], which can lead to inaccurate estimates of degradation damage and mechanical properties.

Transverse transmissions are better for deterioration detection and can provide information on its distribution; however there are some drawbacks. As pointed out in [31], internal seasoning checks can be mistaken for deterioration, moisture content gradients can alter propagation times but be difficult to measure on heavy timbers, and knots close to transducers and oriented with their longitudinal axis in the direction of wave propagation can increase the wave velocity giving misleading data and material representation.

The speed of transmission comparison can be problematic if base line transmission speeds for sound material are not available. If transit speeds of sound material can not be established on site, tables of sound transit times for different species can be consulted. However, not all species have been thoroughly tested to establish the transmission speed of sound members, especially in hardwoods, and variability within species still exists which influences the velocity comparison [33]. Additionally, accurate species identification must be established in order to use these tables reliably and moisture content (of the member of interest comparative to samples used for tabulated speed values) must also be considered. A higher moisture content (up to approximately 30%) will cause an increase in propagation time [3]. Up to 30% moisture content, velocity reduces rapidly with increasing moisture. Above approximately 30% moisture content there is a sharp change in the effect of moisture on velocity and velocity decreases slowly with additional moisture [34]. The effects of preservative treatments, grain orientation and wave propagation direction may also be a factor in assessing velocity comparisons for deterioration detection.

1.4.2 Mechanical Property Prediction

Predictions can be subject to many sources of error including accurate measurements of the member density. Accurate density measurements cannot be measured in a non-destructive manner and therefore must be found by other means and used in combination with stress wave velocity to predict the dynamic modulus. Other methods include tabulated values associated with various species, pilodyn testing, hardness tests, or core sampling. Of these, core sampling and lab evaluation are the most accurate measurements of density. Density established with core samples can be useful, however it will provide only local densities values. Since density of the member may change along the length, multiple samples will be required to establish a mean value for the member as a whole.

Assumptions of the path length of stress wave transmissions can be inaccurate, affecting the dynamic modulus predictions. Path length L in Equation (1.2), is assumed to be a straight line between two locations, but it can be extended due to internal features such as knots, grain deviations or material separations.

Empirical relationships to estimate mechanical properties can have a varying degree of correlation, some quite weak. Independent studies have published conflicting results on correlation values. Investigators and vendors at times overstate the correlation between measured quantities and the strength of the tested structural component and that strong correlations are questionable because of the indirect link between the parameters and the strength [24]. Property estimate correlations are further reduced due to compounded correlations. Typically a dynamic modulus is calculated from stress wave velocity measurements, which has its own correlation value. After estimating the dynamic modulus, empirical relationships with varying correlation coefficients are used to predict strength values. The total correlation between the stress wave parameter and the predicted strength value would then be the product of the two correlation coefficients, a reduction in total correlation since correlation coefficients are less than one (excluding perfect relationships).

References

1. Anthony, R.W., Bodig, J., Phillips, G.E. and Brooks, R.T. Longitudinal nondestructive evaluation of new utility wood poles. EPRI TR-100864, Vol. 1, Project 3078-01, 1992.
2. Kaiserlik, J. Nondestructive testing methods to predict effect of degradation on wood: A critical assessment. GTR-FPL-019, Gen. Tech. Rep. FPL-19. Madison, WI: US Department of Agriculture, Forest Service, Forest Products Laboratory, 53 pp., 1978.
3. Ross, R.J., Brashaw, B.K., Wang, X., White, R.H., and Pellerin, R.F. *Wood and Timber: Condition Assessment Manual*. Forest Products Society, Madison, WI, 73 pp., 2004.
4. Beall, F.C. Overview of the use of ultrasonic technologies in research on wood properties. *Wood Science and Technology*, Vol. 36, pp. 197–212, 2002.
5. Bozhang, S. and Pellerin, R.F. Nondestructive evaluation of the degree of deterioration in wood: Stress wave frequency spectrum analysis. In: *Proceedings of 10th International Symposium on Nondestructive Testing of Wood*. Presses Polytechniques et Universitaires Romandes, Lausanne, Switzerland, pp. 99–115, 1996.
6. Lempriere, B.M. *Ultrasound and Elastic Waves: Frequently Asked Questions*. Academic Press, San Diego, 2002.
7. Emerson, R.N., Pollock, D.G., McLean, D.I., Fridley, K.J., Ross, R.J., and Pellerin, R.E. Nondestructive testing of large bridge timbers. In: *Proceedings of 11th International Symposium on Nondestructive Testing of Wood*. Madison, WI, pp. 175–184, 1999.
8. Bucar, V. *Nondestructive Characterization and Imaging of Wood*. Springer, Berlin, pp. 181–213, 2003.

9. Hasenstab, A. Integritätsprüfung von Holz mit dem zerstörungsfreien Ultraschall-echoverfahren. Doctoral Dissertation Bundesanstalt für Materialforschung und -prüfung BAM-Dissertationsreihe, Band 16, Berlin, 2006.
10. Hasenstab, A., Hillemeier, B., and Krause, M., Defect localisation in wood with low frequency ultrasonic echo technique. In: *Proceedings of 14th International Symposium on Nondestructive Testing of Wood*, Hannover, Germany, 2005.
11. Hasenstab, A. 2007. Ultraschall-Echo zur Ortung von Rissen in Brettschichtholz (BSH). Tagungsband der DGZfP-Jahrestagung, Vortrag 39, 14.–16. May 2007, Fürth, 2007.
12. Maack, S. and Krause, M. Diagnostic investigations of wooden structures using ultrasonic-echo technique. In: *Proceedings of 1st International RILEM Symposium On-Site Assessment of Concrete, Masonry and Timber Structures*, Varenna, Italy, 2008.
13. Hasenstab, A., Osterloh, K., and Krause, M. Testing of wooden construction elements with ultrasonic echo technique and X-ray. In: *Proceedings of 9th European Conference on NDT*, September 25–29, 2006, Berlin Deutsche Gesellschaft für Zerstörungsfreie Prüfung e.V. BB 103-CD Th.2.4.1, 2006.
14. ACSYS Acoustic Control Systems, Moscow, Russia, online: http://www.acsys.ru, 2005.
15. Newman, A. *Structural Renovation of Buildings: Methods, Details, and Design Examples*. McGraw-Hill, New York, 2001.
16. Dinwoodie, J.M. *Timber: Its Nature and Behavior*, 2nd ed. E&FN Spon, New York, 2000.
17. Zombori, B. "In situ" nondestructive testing of built in wooden members. NDT.net 6.3 http://www.ndt.net/article/v06n03/skatter/skatter.htm, 2001.
18. Ross, R.J., DeGroot, R.C., Nelson, W.J., and Lebow, P.K. The relationship between stress wave transmission characteristics and the compressive strength of biologically degraded wood. *Forest Products Journal*, Vol. 47(5), pp. 89–93, 1997.
19. Bucar, V. and Bohnke, I. Factors affecting ultrasonic measurements in solid wood. *Ultrasonics*, Vol. 32(5), pp. 385–388, 1994.
20. Bucur, V. Techniques for high resolution imaging of wood structure: A review. *Measurement Science and Technology*, Vol. 14, pp. 91–98, 2003.
21. Raj, B., Jayakumar, T., and Thavasimuthu, M., *Practical Non-Destructive Testing*, 2nd ed. Narosa Publishing House, New Delhi, pp. 77–99, 2002.
22. ASTM E494-95. Standard Practice from Measuring Ultrasonic Velocity in Materials, ASTM International.
23. Peterson, M.L., Downs III, J. and Gutkowski, R.M. Nondestructive inspection of timber bridge structures. In: *Proceedings of Nondestructive Evaluation of Civil Structures and Materials*, University of Colorado, Boulder, CO, pp. 55–67, 1996.
24. Anthony, R.W. and Bodig, J., Application of stress wave-based nondestructive evaluation to wood. In: *Proceedings of Nondestructive Evaluation of Civil Structures and Materials*. University of Colorado, Boulder, CO, pp. 257–276, 1990.
25. Ross, R.J., and Pellerin, R.F. Nondestructive testing for assessing wood members in structures: A review. Gen. Tech. Rep. FPL-GTR-70. US Department of Agriculture, Madison, WI, Forest Service, Forest Products Laboratory, pp. 27–48, 1991.
26. Alexander, A.M. Accuracy of predicting in situ compressive strength of deteriorated concrete seawall by NDT methods. In: *Proceedings of Nondestructive Evaluation of Civil Structures and Materials*. University of Colorado, Boulder, CO, pp. 68–82, 1992.
27. Galligan, W.L. and Courteau, R.W. Measurement of elasticity of lumber with longitudinal stress waves and the piezoelectric effect of wood. In: *Proceedings of 2nd International Symposium on Nondestructive Testing of Wood*. Washington State University, National Science Foundation, Spokane, WA, pp. 223–244, 1965.
28. Wang, X., Ross, R.J., Erickson, J.R., Forsman, J.W., McGinnis, G.D., and De Groot, R.C. Nondestructive methods of evaluating quality of wood in preservative-treated piles. Research Note. FPL-RN-0274. US Department of Agriculture, Forest Service, Forest Products Laboratory, Madison, WI, 2000.
29. Feio, A.O., Machado, J.S., and Lourenco, P.B. Parallel to the grain behavior and NDT correlations for chestnut wood (Castanea Sativa Mill). In: *Proceedings of Conservation of Historic Wooden Structures*. Florence, Italy, pp. 294–303, 2005.

30. Bucar, B. and Feeney, F. Attenuation of ultrasound in solid wood. *Ultrasonics*, Vol. 30(2), pp. 76–81, 1992.
31. Ceccotti, A. and Togni, M. NDT on ancient timber beams: Assessment of strength/stiffness properties combining visual and instrumental methods. In: *Proceedings of 10th International Symposium on Nondestructive Testing of Wood*. Presses Polytechniques et Universitaires Romandes, Lausanne, Switzerland, pp. 379–388, 1996.
32. Pellerin, R.F., DeGroot, R.C., and Esenther, G.R. Nondestructive stress wave measurements of decay and termite attack in experimental wood units. In: *Proceedings of 5th Nondestructive Testing of Wood Symposium*. Washington State University, USDA Forest Products Laboratory, Pullman, WA, pp. 319–352, 1985.
33. Ross, R.J., Soltis, L.A., and Otton, P. Role of nondestructive evaluation in the inspection and repair of the USS constitution. In: *Proceedings of 11th International Symposium on Nondestructive Testing of Wood*. Forest Product Society, Madison, WI, pp. 145–152, 1999.
34. Sakai, H., Minamisawa, A., and Takagi, K. Effect of moisture content on ultrasonic velocity and attenuation in woods. *Ultrasonics*, Vol. 28(6), pp. 382–385, 1990.

Chapter 2
Ground Penetrating Radar

Mehdi Sbartaï

Abstract Ground Penetrating Radar (GPR) called also Radar (Radio Detecting and Ranging) is an application of electromagnetic (EM) waves. It has been developed in the last decades for detection, distance measurement, defects and anomaly localization, and characterization of dielectric materials such as soil, concrete, masonry and wood. The frequency range of the technique varies between 100 MHz to few GHz. It is a nondestructive technique based on the transmission of EM energy according to electromagnetism theory.

Key words: GPR, radar, permittivity, defect detection, moisture evaluation

2.1 Background

The EM waves propagate through nonmagnetic materials. This propagation is described by an energy movement depending on the relation between electric and magnetic fields. In the case of air propagation, the two fields are perpendicular and their amplitudes are in a constant ratio. However, the propagation of EM waves in nonmagnetic materials is governed by the permittivity of the medium. This EM property will be defined in the next section.

EM theory has been described by Maxwell in 1861 according to the four following equations named "Maxwell equations":

$$\begin{aligned} \operatorname{rot}\mathbf{E} &= -\mu\frac{\partial\mathbf{H}}{\partial t} \\ \operatorname{rot}\mathbf{H} &= \mathbf{J}_c + \mathbf{J}_d \\ \mathbf{J}_c &= \sigma\times\mathbf{E} \\ \mathbf{J}_d &= \varepsilon\frac{\partial\mathbf{E}}{\partial t} \end{aligned} \tag{2.1}$$

B. Kasal and T. Tannert (eds.), In Situ Assessment of Structural Timber, 25–37.

The combination of Maxwell equations leads to the wave equations of the electric and magnetic fields of direction x:

$$\Delta E(x,t) = \mu\varepsilon\left[\frac{\partial^2 E}{\partial t^2}\right]$$

$$\Delta H(x,t) = \mu\varepsilon\left[\frac{\partial^2 H}{\partial t^2}\right] \quad (2.2)$$

where **E** denotes the electric field [V/m], ε is the permittivity of the medium [F/m], **H** is the magnetic field [T], μ is the magnetic permeability of the medium [H/m], $\mathbf{J}_c$ is the current density of conduction [A/m^2], $\mathbf{J}_d$ is the current density of displacement [A/m^2], x is the travelled distance, and t is time.

In the case of plane wave equation, which is generally taken for GPR, the solution of wave equation of the electric field can be assumed as follows:

$$E(x) = E_0 e^{-\gamma x} \quad (2.3)$$

where E_0 is the amplitude of the generated electric field, and γ is the complex propagation constant

$$\gamma = i\omega\sqrt{\varepsilon\mu} = \alpha - i\beta \quad (2.4)$$

where the real part is the attenuation constant

$$\alpha = \omega\left[\frac{\mu_0\varepsilon_0\varepsilon'}{2}\left(\sqrt{1+\left(\frac{\varepsilon''}{\varepsilon'}\right)^2} - 1\right)\right]^{1/2} \quad (2.5)$$

and the imaginary part is the phase constant

$$\beta = \omega\left[\frac{\mu_0\varepsilon_0\varepsilon'}{2}\left(\sqrt{1+\left(\frac{\varepsilon''}{\varepsilon'}\right)^2} + 1\right)\right]^{1/2} \quad (2.6)$$

where ω is the angular frequency; ε_r' and ε_r'' are the real and imaginary part of the material permittivity, respectively; μ_0 and ε_0 are the electromagnetic permeability and permittivity of free space, respectively. For slightly conducting media, typically for wood with low moisture content, phase and attenuation can be simplified as follows:

$$\beta = \omega\sqrt{\mu_0\varepsilon_0\varepsilon'}$$

$$\alpha = \frac{\omega\varepsilon''}{2}\sqrt{\frac{\mu_0}{\varepsilon'}} \quad (2.7)$$

The propagation velocity of EM waves in wood material can be written as

$$v = \frac{\omega}{\beta} = \frac{1}{\sqrt{\mu_0 \varepsilon_0 \varepsilon'}} = \frac{c}{\sqrt{\varepsilon'}} \tag{2.8}$$

c is the propagation velocity of EM waves in free space:

$$c = \frac{1}{\sqrt{\mu_0 \varepsilon_0}} \approx 3 \times 10^8 \text{ m/s} \tag{2.9}$$

The previous equations show that propagation velocity and attenuation of EM waves are directly related to the permittivity of dielectric materials such as wood. Therefore, the permittivity of wood must be studied to understand GPR wave propagation in wood material.

2.1.1 Dielectric Permittivity of Wood

Wood is a natural and nonmetallic material. It is considered a dielectric material and a slightly conducting medium permitting the propagation of EM waves. It is characterized by a dielectric permittivity which can be written in terms of real and imaginary parts as

$$\varepsilon = \varepsilon' - i\varepsilon'' \tag{2.10}$$

The first property is defined as the capacity of the material to store EM energy (electrical polarization phenomena) and the second represents the loss of energy due to absorption mechanisms (electrical conduction and dielectric relaxation). The real part of the complex permittivity is known as the dielectric constant (ε) and imaginary part is the loss factor (ε''). The ratio ($\varepsilon''/\varepsilon'$) is named the loss tangent and written as ($\tan\delta$).

Generally, the measurement of the relative permittivity of wood is based on microwave technique which involves the propagation of EM waves from antenna at frequencies varying from 3 to 300 GHz, approximately. It can be used in two configurations: transmission or reflection as can be seen in Figure 2.1.

In the first mode, the material is located between two antennas (one for transmitting microwave and the other for reception). In the second mode, both antennas are located in front of the material. Usually, the following components are required for experimental device:

- Two horn antennas (transmitting and receiving);
- Generator of electromagnetic wave at relatively high frequency and low power;
- Data acquisition system.

It is well known that the permittivity of wood depends on several parameters such as moisture content, density, temperature, fiber direction and the applied frequency. This EM property of wood has been studied by several researchers on various species (different wood origin) having various moisture content and anisotropic directions, at different temperature and frequency [1–5].

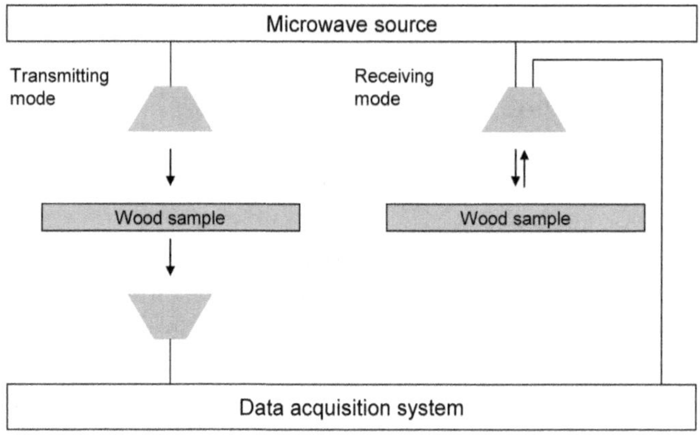

Fig. 2.1 Schematic principle of microwave system for wood permittivity measurement

Wood can be considering as an orthotropic material, the dielectric properties can be written in the form of tensor. If the direction of the electrical field coincide with that of the principal anisotropic directions (L, R, and T), the dielectric constant and the loss tangent are defined as follows [6]:

$$[\varepsilon'] = \begin{bmatrix} \varepsilon'_{11} & & \\ & \varepsilon'_{22} & \\ & & \varepsilon'_{33} \end{bmatrix}$$

$$[\tan\delta] = \begin{bmatrix} \tan\delta_{11} & & \\ & \tan\delta_{22} & \\ & & \tan\delta_{33} \end{bmatrix} \tag{2.11}$$

However, as demonstrated in [2], the dielectric properties in the directions R and T are almost similar and especially the loss tangent. Table 2.1 reports measurement of the dielectric constant and the loss factor of Douglas fir at 8.53 GHz at various moisture content and fiber directions. It can be noted an increase of the dielectric constant and loss factor with respect to moisture due to polarization mechanisms. In wood material, dipole polarization is caused by free and bound water molecules, which have permanent dipole. Accumulation of charges ions at the interfaces between different zones of the cell wall due to the application of electric field at low frequencies is called interfacial polarization that induces loss of EM energy. These phenomena are increased with respect to water molecule in wood.

Considering measurement at a constant temperature and frequency, the dielectric properties of wood depend also on the density of the material. This dependence is available in the literature, i.e. Torgovnikov [7] demonstrates the relation between dielectric properties of wood and density at various moisture contents as can be seen in Figure 2.2. From this figure, it can be concluded that both EM properties of wood are related linearly to the density and moisture content.

Table 2.1 Effect of moisture content and fiber direction on the dielectric properties of Douglas fir at 8.53 GHz [2]

Moisture content (%)	Direction L ε'	Direction L $\tan\delta$	Direction R ε'	Direction R $\tan\delta$	Direction T ε'	Direction T $\tan\delta$
7	1.9	0.14	1.7	0.07	1.8	0.09
10	2.1	0.17	1.9	0.10	1.9	0.11
12	2.6	0.22	1.9	0.11	2.1	0.13
16	2.9	0.26	2.1	0.18	2.3	0.21
22	4.2	0.45	2.6	0.26	3.0	0.25

Dielectric properties of wood, which govern EM wave's propagation, depend on moisture content, density, and fiber direction, at fixed frequency and temperature. Therefore, EM waves can be applied to the nondestructive evaluation of wood physical properties. In this context, many instruments based on the transmission of EM energy at microwave frequencies have been developed in order to control wood material such as moisture and density evaluation [9–12] as well as knots detection [13] and grain angle measurement [14]. However, a few of these techniques can be used for on site evaluation of timber structures. For this reason, this chapter is restricted to GPR applications for the on-site evaluation of timber structures.

2.2 Equipment

GPR has been used firstly for military applications, i.e. detection of planes during the Second World War. The first application of GPR as a nondestructive technique for the reconstruction of geophysical structures appeared in 1950. From the on, the application of GPR technology has increased and a new generation of acquisition system and antennas have been developed.

Equipment is composed of an EM waves pulse generator, a power supply, transmitting (T) and receiving (R) antennas, and a computer for signals processing. The measurement is based on the radiation of EM energy using dipole or horn couple of T-R antennas. This energy propagates through the material, attenuated, reflected, and refracted on any interface presenting contrasts of permittivity (Figure 2.3). The analysis of the recorded signal (i.e. velocity, attenuation) can give physical and/or geometrical information about the material.

- S_1: signal of the direct wave propagating directly from the transmitter to the receiver;
- S_2: reflected signal on heterogeneity;
- S_3: reflected signal on interface between two materials having different permittivity;
- S_4: reflected signal on the interface between material 2 and air;
- S_5: refracted signal.

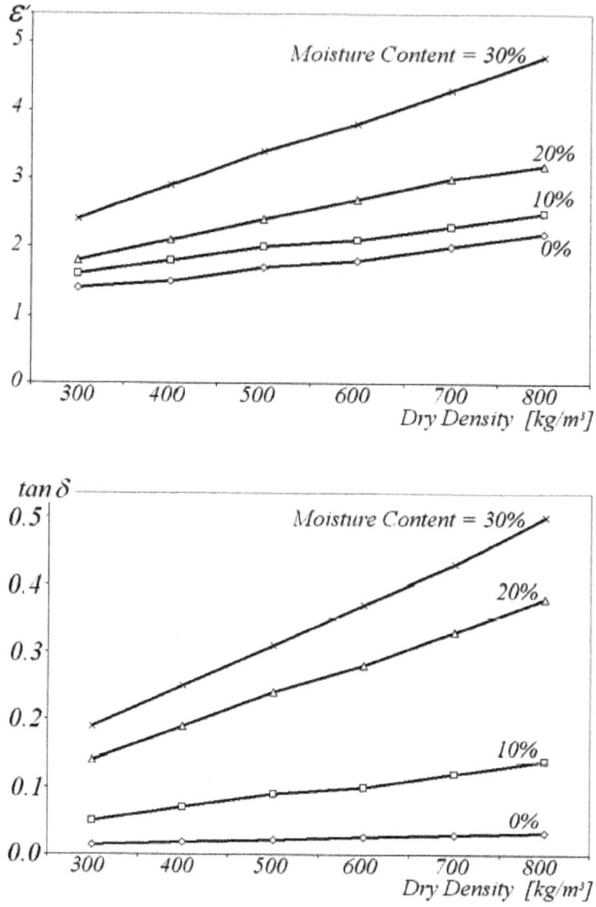

Fig. 2.2 Effect of moisture content and density on the dielectric properties [8]

The most important advantage of GPR technique is its ability to investigate a large surface within relatively short time. The system generates a series of short pulses that travel through the material and back-scatter during a few nanoseconds. The technique can then be implemented quickly. In addition, the newly developed system can generate EM pulses at speed higher than 64 scans per second. Generally, GPR measurements consist in recording a profile by moving the antennas on the tested structure along a linear direction. The system is able to record a "trace" or (A-scan) every centimeter of the profile. Figure 2.4 presents a GPR trace recorded on dielectric material with several defects. Different reflections can be clearly observed on the recorded trace, which indicate the presence of EM contrast (change of the material permittivity at the interface between damaged and non-damaged zones).

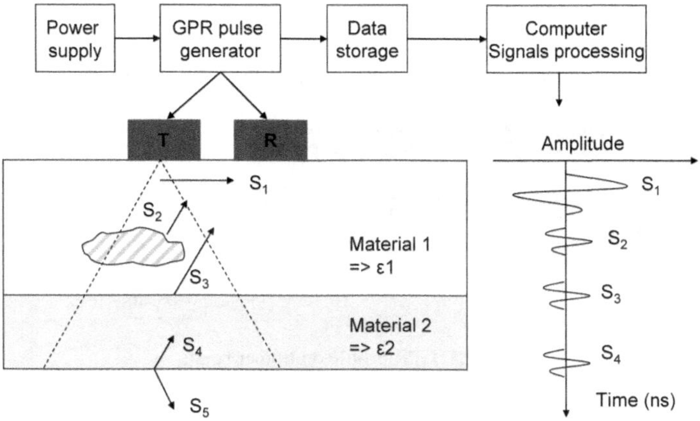

Fig. 2.3 Schematic principle of GPR measurement using ground-coupled antennas

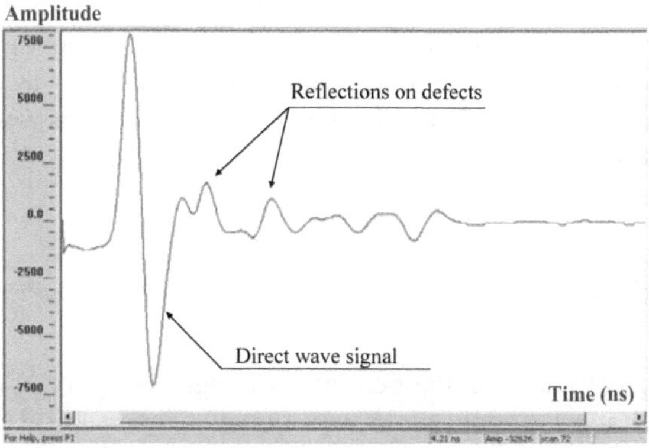

Fig. 2.4 GPR trace "A-scan" recorded on dielectric material with several defects

The juxtaposition of the recorded traces can be used to reconstitute a 2D image (position of the antennas on the material with respect to time position of the different reflections in depth). This image is generally called "radargram" as can be seen in . In this figure the reflection generated by the contrast of permittivity between wood material and air (opposite side of the sample) can be clearly observed.

As the recorded trace is amplitude versus time, the depth of reflections can approximated by the following simplified equation:

$$d = \frac{v \times t}{2} \tag{2.12}$$

where d is the depth (m), v is the propagation velocity (m/ns), and t is the recorded time position of the signal (ns).

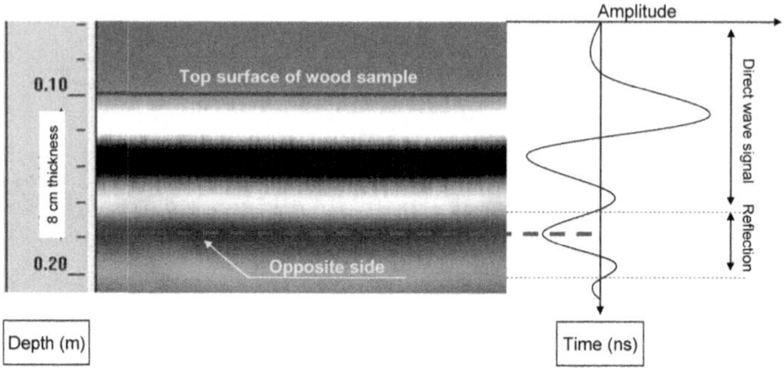

Fig. 2.5 GPR profile on timber beam

Regarding this equation, the propagation velocity must be calculated or evaluated. This velocity theoretically depends on the dielectric constant of wood material. If the dielectric constant is known, the velocity can be calculated. However, on site, this property is difficult to be measured. In addition, as reported previously, ε' varies depending on the physical properties of the materiel such as moisture, density as well as fiber direction. Generally, the dielectric constant is estimated by GPR calibration i.e. knowing the thickness of the tested element.

2.3 Applications

GPR has been largely used in the field of civil engineering for bridges and roads inspection. Generally, the applications were limited to geometrical characterization (thickness evaluation of concrete or masonry elements and depth measurement of embedded reinforcement in concrete structures) [15]. In addition, GPR has been used for defects detection such as cracks as well as delamination [16]. Recently, this technology has been applied successfully for physical characterization of concrete such as moisture and chlorides evaluation [17–19]. Some researchers have used GPR for qualitative detection of moisture on masonry structures [20].

However, limited applications are available regarding timber structures assessment using GPR technology. This is mainly due to the difficulty of interpreting GPR data and to the fact that the results are generally qualitative information. The literature concerning this field of application is poor and deals with the application of GPR for timber bridge inspections. This research field has been mainly implemented in the USA and Australia. This is because, for both countries, a significant proportion of bridges are designed with timber elements such as decks, piles or girders. The major applications concern the detection of defects due to termite and fungal attack which result in the reduction of bridge load capacity.

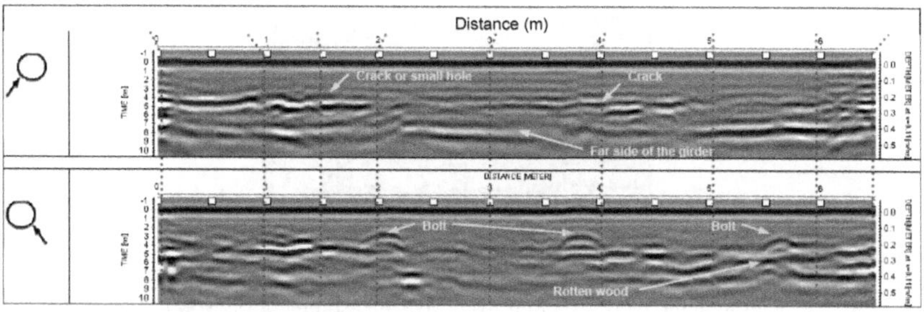

Fig. 2.6 GPR profiles on girder of a timber bridge [21]

2.3.1 Detection

In this field of GPR application, Muller et al. [21] reported that this technique was well suited to inspection of timber bridges for piping and rotting defects. The authors used high frequency ground coupled dipole antenna at 1.2 GHz. The tests were carried out on girder from a bridge after its demolition. Several radargrams have been recorded along the girder on two different positions. The validation of radar data was confirmed by cutting up the girder at one meter intervals and comparing radar detections with defects that were found as can be seen in Figure 2.6. On the two radargrams, cracks and rotten wood are clearly identified. These observations are due essentially to the reflections on defects leading to the modification of the permittivity of the material between damaged and non-damaged wood.

Hernandez and Duwadi [22] have developed a micropower impulse radar (MIR) imaging system for the NDE of timber structures. The MIR was composed of horn antenna (T-R), portable computer with PCMCIA card, and a generator of ultra-wideband pulses having bandwidth of 4 GHz, approximately. The initial investigations were performed on $122 \times 42 \times 22$ cm Douglas fir glue-laminated beam samples. One sample was penetrated with various known size voids. The tests consist of recording linear scans and the results show that the developed prototype is able to detect a variety of voids in the tested wood sample as seen in Figure 2.7. The authors emphasized that the integration of an RF-based position encoding mechanism would be useful to the MIR. Moreover, further tests of various specimens and the optimizing of the image-reconstruction are needed for enhancing this application.

In the field of historical structures inspection, GPR was used to detect timber beams and to evaluate the type and dimension of their connection to bearing wall by 3D reconstruction imaging [23]. They implemented some GPR investigations for detecting small beams supporting structure in timber floors as can be seen in Figure 2.6. The investigations were implemented using GPR connected to 1 GHz antenna by recording parallel profiles orthogonal to the beam direction. The authors reported that 3D GPR reconstruction is a powerful technique for the detection and dimension evaluation of hidden structure of timber floors.

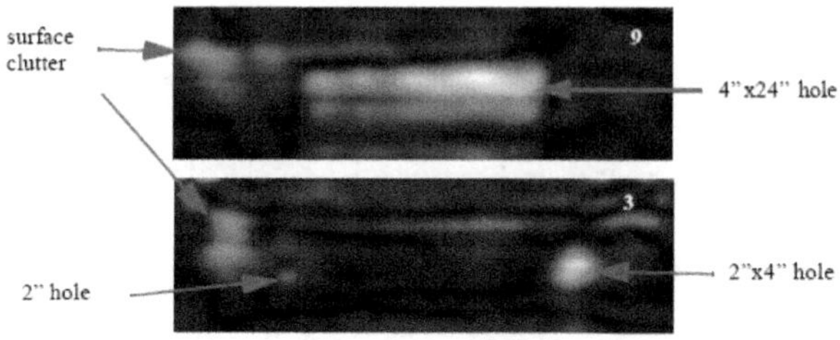

Fig. 2.7 Tomographic reconstruction from MIR measurement on wood sample [22]

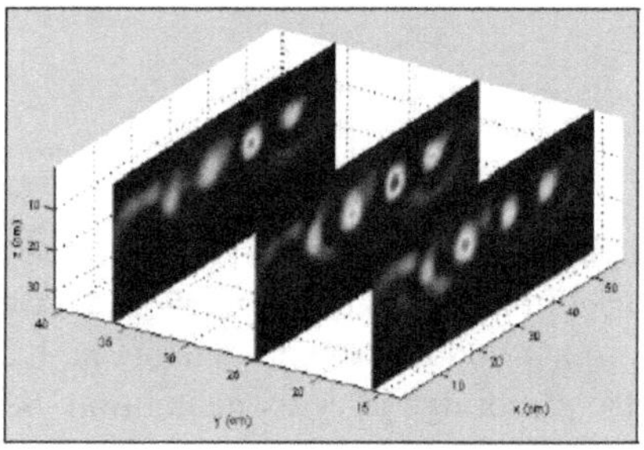

Fig. 2.8 3D detection of beam in timber floor structure [23]

2.3.2 Moisture Evaluation

EM wave propagation is governed by moisture content of the material. This is because the dielectric constant of water is higher compared with that of wood which is equal to approximately 2 at dry state. Then, higher moisture in wood increases the dielectric constant and consequently decreases the propagation velocity of radar waves.

In this context, Lualdi et al. [23] have conducted laboratory tests with the goal of studying the sensitivity of radar signals to moisture variation in wood material. Radar measurements were carried out using GSSI SIR 2000 system connected to 1.5 GHz antenna on three wood samples (spruce, pine, and spruce glue-laminated timber) as seen in Figure 2.9. The samples having dimensions of $80 \times 20 \times 9$ cm were placed vertically in water to create moisture gradient in the longitudinal direction. The antenna was placed on the sample at three different points: A (high moisture), B (medium moisture), and C (low moisture). At each point, longitudinal and transverse measurements were taken.

The measured propagation time of the direct signal was affected by the moisture content of wood (Figure 2.10). Moreover, this radar feature seems to be also affected by the direction of fiber. The time propagation measured perpendicular to fiber was higher compared to parallel direction. This indicates that the propagation velocity decreased in transversal direction compared to longitudinal direction. The results confirm the effect of water and fiber direction on the dielectric constant of

Fig. 2.9 GPR measurement on a spruce glue-laminated sample using ground-coupled antenna at 1.5 GHz [24]

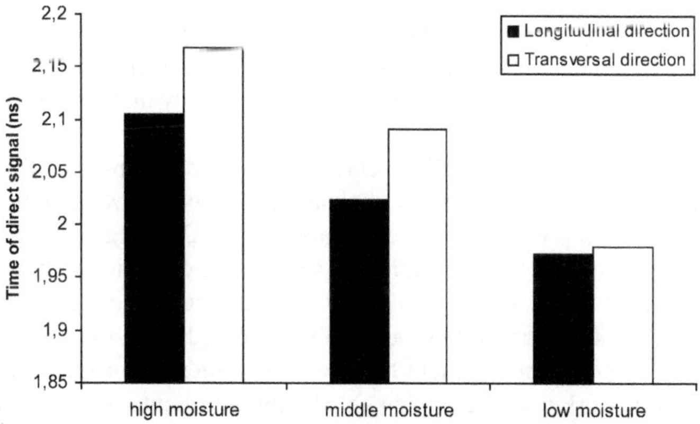

Fig. 2.10 Effect of moisture and fiber direction on the propagation time of GPR waves [24]

wood. Therefore, GPR system can be applied to physical characterization of timber structures such as moisture evaluation.

2.4 Limitations

GPR is a powerful technique for rapid on-site investigations. The system is light and portable and the measurement is easy. This technique offers the possibility of being applied to defects detection and to moisture content and fiber direction evaluation. However, the technique requires skilled operator to interpret the results. In addition, the data processing is time consuming.

GPR has been applied successfully to defects detection in timber girder and also for detecting timber beam in masonry structures. However, the depth information given by GPR can be disturbed by the variation of moisture or density in wood.

Therefore, the system must be calibrated to enhance the evaluation of the depth of defects, cracks or connections. Another limitation pertains to resolution of the system, which is directly related to the applied frequency. The generation of GPR waves at high frequency results in a small wavelength. If the wavelength is larger than the dimension of the heterogeneity, the resolution can be very low and in some cases the detection is impossible.

GPR penetration depth can also be limited in case of material with high permittivity. In fact, increasing permittivity indicates an increase of absorption and conduction. Then, GPR waves are rapidly attenuated, and the penetration depth decreases. The penetration depth is governed by moisture content and density of wood. It also depends on used frequency. Low frequencies have large wavelengths and high capacity of penetration compared with high frequencies. For example, at 1.5 GHz frequency, the penetration depth of GPR waves in material with high moisture content is less than 40 cm.

The detection capacity of GPR is also affected by the permittivity contrast between the material and the defect. If the two media have similar EM property, the defect cannot be detected. However, the detection capacity increases with respect to the contrast.

Moisture evaluation using GPR is currently under development. The results of research works show only qualitative evaluation of wood moisture content. The research consists of analysing the effect of moisture on GPR signal features such as velocity or attenuation. This application on site is more complicated because timber properties change in the structure (density, fiber direction, heterogeneity, defects, etc.). In addition, these properties can also vary according to their position within the thickness of the timber element leading to a gradient of properties. Therefore, further research work must be implemented with taking into account the variation of timber properties in order to consider the prediction of moisture content.

References

1. Hearmon, R.F.S. and Burcham, J.N. The dielectric properties of wood. For. Prod. Res. Special Report No 8, Dep. of Sci. and Industry, London, 25 pp, 1954.
2. James, W.L. and Hamill, D.W. Dielectric properties of Douglas fir measured at microwave frequencies. *Forest Products Journal*, Vol. 15(2), pp. 51–56, 1965.
3. James, W.L. Dielectric properties of wood and hardboard: variation with frequency, moisture content and grain orientation. USDA for Serv. Res. Pap. FPL-245, Madison, WI, 1975.
4. Peyskens, E., de Pourq, M., Stevens, M., and Schalck, J. Dielectric properties of softwood species at microwave frequencies. *Wood Science Technology*, Vol. 18, pp. 267–280, 1984.
5. Dubey, Y.M. and Deorani, S.C. Dielectric properties of coniferous timbers at microwave frequencies. *Journal of Indian Acad. of Wood Sci.*, Vol. 1, pp. 77–82, 1997.
6. Bucur, V. and Timell, T.E. *Nondestructive Characterization and Imaging of Wood*, Springer Series in Wood Science, Springer, Berlin, pp. 128–129, 2003.
7. Torgovnikov, G.I. *Dielectric Properties of Wood and Wood Based Materials*. Springer, Berlin, 196 pp., 1990.
8. Stalnaker, J.J. and Harris, E.C. *Structural Design in Wood*, 2nd ed. Kluwer Academic Publishers, Boston, 2002.

9. Johansson, J., Hagman, O., and Oja, J. Predicting moisture content and density of Scots pine by microwave scanning of sawn timber. *Computers and Electronics in Agriculture*, Vol. 41, pp. 85–90, 2003.
10. Johansson, J., Hagman, O., and Fjellner, B.A. Predicting moisture content and density distribution of Scots pine by microwave scanning of sawn timber. *J. Wood Sci.*, Vol. 49, pp. 312–316, 2003.
11. Hansson, L., Lundgren, N. Antti, A.L. and Hagman, O. Microwave penetration in wood using imaging sensor, *Measurement*, Vol. 38, pp. 15–20, 2005.
12. Lundgren, N., Hagman, O., and Johansson, J. Predicting moisture content and density distribution of Scots pine by microwave scanning of sawn timber II: Evaluation of models generated on a pixel level. *J. Wood Sci.*, Vol. 52, pp. 39–43, 2006.
13. Baradit, E., Aedo, R., and Correa, J. Knots detection in wood using microwaves. *Wood Science Technology*, Vol. 40, pp. 118–123, 2006.
14. Schajer, G.S. and Orhan, F.B. Measurement of wood grain angle, moisture content and density using microwaves. *Holz als Roh- und Werkstoff*, Vol. 64, pp. 483–490, 2006.
15. Mallinson, P. Surface penetrating radar as an NDT tool. In: *Proceedings 36th Annual British Conference on NDT*, Cheltenham, UK, 1997.
16. Halabe, U.B, Bhandarkar, V., Chen, H-L., and Sami, Z. Detection of sub-surface anomalies in concrete bridge decks using ground penetrating radar. *ACI Materials Journal*, Vol. 94(5), pp. 396–408, 1997.
17. Sbartaï, Z.M., Laurens, S., Balayssac, J.-P., Ballivy, G., and Arliguie, G. Effect of concrete moisture on radar signal amplitude. *American Concrete Institute – Materials*, Vol. 103(6), pp. 419–426, 2006.
18. Sbartaï, Z.M., Laurens, S., Rhazi, J., and Balayssac, J.-P. Nondestructive evaluation of concrete structures using geophysical methods: Correlation between electrical resistivity and radar measurements. *Journal of Applied Geophysics*, Vol. 62(4), pp. 361–374, 2007.
19. Klysz, G. and Balayssac, J.-P. Determination of volumetric water content of concrete using ground-penetrating radar. *Cement and Concrete Research*, Vol. 37(8), pp. 1164–1171, August 2007.
20. Maierhofer, C., Wöstmann, J., Trela, C., and Röllig, M. Investigation of moisture content and distribution with radar and active thermography. In: *Proceedings of RILEM International Conference, SACOMATIS*, Varenna, Italy, 2008.
21. Muller, W. Timber girder inspection using ground penetrating radar. In: *Proceedings NDT-CE International Conference*, 2003.
22. Hernandez, J.E. and Duwadi, P.E. An impulse radar tomographic imaging system for NDE of timber structures. TRB Paper No. 00-0558, 2000.
23. Lualdi, M., Zanzi, L., and Binda, L. Acquisition and processing requirements for high quality 3D reconstructions from GPR investigations. In: *Proceedings NDT-CE International Conference*, 2003.
24. Sbartaï, Z.M., Bos, F., and Breysse, D. NDE of timber structures using GPR. In: *Structural Faults & Repair*, 15–17 June 2010.

Chapter 3
Radiography

Bohumil Kasal, Gretchen Lear and Ron Anthony

Abstract Radiography evolved from laboratory testing into in-situ applicable methods that utilize low-energy gamma X-ray sources. Due to the low attenuation properties of wood and difference in attenuation of different material, the radiography is relatively easily applicable to timber. Parameters such as energy level, time of exposure and distance between the source, object and imaging plate permit identification of early and late wood. Quantitative radiography can be used to extract dimensions, deformations or even strains. One of the challenges is the collapse of a three dimensional object into two dimensions. Access must available from opposite sites of the member.

Key words: X-ray, gamma rays, quantitative radiography, radioscopy, attenuation, relative attenuation, sharpness, source, digital imaging, wavelength, spectrum, cracks

3.1 Background

Radiography uses penetrating radiation to depict the internal structure of members. A radiation source is used to emit a beam of radiation directed towards an object of interest. Objects under investigation will have varying absorption of radiation based on the material density and thicknesses. Opposite the source of radiation and behind the object of interest is a radiation sensitive film or recording medium that produces images (see Figure 3.1). This noninvasive procedure allows for extensive investigation into issues such as structure composition, hidden internal materials and flaws and the state of preservation, which at times cannot be gained by other means.

Digital imaging systems and digital radioscopy have seen advancement because of their traditional use in security, bomb and drug detection, and forensics as well as industrial non-destructive testing for quality assurance. The advancement seen from these commercial uses have lead to the development of highly portable and user

B. Kasal and T. Tannert (eds.), In Situ Assessment of Structural Timber, 39–50.

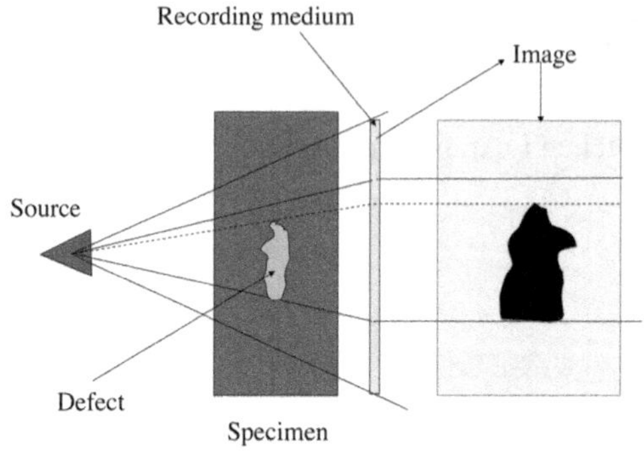

Fig. 3.1 General arrangement for radiographic imaging

friendly systems that can be run either on battery or with common AC/DC adaptors depending on the system. Laptop computers are generally used with these systems and allow for nearly instant image viewing and manipulation. Such advancements have led the technology to be readily applied to in situ investigations.

3.2 Equipment

3.2.1 Technology

Penetrating radiation used in radiography is generated from several sources, most commonly electrons, neutrons, gamma rays and X-rays. Electrons used in radiography are produced either by radioactive decay or high-energy X-ray impact on a heavy metal. Because of the strong absorption of electrons by all materials, the penetration power is limited and restricts the usefulness of this radiation source to thin, low density material [1].

Neutrons are produced by linear accelerators or nuclear reactors and are prone to absorption by organic material. Neutrons are rarely used for in situ investigation for multiple reasons including the limited access to linear accelerators or nuclear reactor sources, the expense of equipment and operation, issues related to the difficulty of on-site equipment set up, limitation to small areas of investigation due to the narrowness of the radiation beam produced, poor reaction of neutrons with film, and the possibility of objects becoming radioactive after investigation [1]. These drawbacks limit the use of electron and neutrons as radiation sources for in situ investigation and the remainder of the discussion will focus on gamma and X-ray radiation sources.

Gamma rays and X-rays are the most commonly used sources of radiation used for radiography. They are short wavelength electromagnetic radiations which are physically the same but differ in the way they are produced. Both travel in straight lines at the speed of light, can be diffracted but not deflected, and are unaffected by electrical or magnetic fields. The rays penetrate matter, the degree of which is dependent on material type, density and thickness as well as the radiation energy [1, 2].

3.2.2 Gamma Rays

Gamma rays are emitted during the radioactive decay of unstable isotopes, each having a characteristic energy and intensity for the radiation it emits. The high energy levels of the gamma rays create substantial penetration capabilities. Isotope energy remains constant; however, the intensity decays with time as indicated by the half life. The wavelengths of the radiation produced by the gamma sources are distinct and limited as opposed to X-rays which have a broad wavelength spectrum.

Although gamma and X-rays are physically the same, the production differences have a distinct effect on gamma rays use for in situ evaluation. Gamma rays have the advantage of a portable nature since radioactive isotopes do not require the external energy or cooling sources that X-ray generators do. The elimination of external power, as well as the reduction and compactness of the equipment, make the method more mobile and less expensive.

The advantages of a radiation source with no external power are limited. The radioactive isotopes continually generate radiation and require special containers lined with lead for storage to protect against the harmful effects of the radiation on living tissue. In addition, when use of the source is needed it must be removed from the storage container by means of a remote controlled mechanical device. The source also has a limited life span as it loses its intensity over time, depending on the half-life of the isotope in use, and the high energy radiation of gamma rays cannot be controlled, resulting in poorer quality imaging with lower contrast than X-rays [1].

3.2.3 X-Rays

X-rays are produced when high-speed electrons impact matter. Energy is lost upon impact and a small fraction is converted into short wavelength radiation. The remaining impact energy is mostly converted to heat. The X-ray spectrum is comprised of two underlying spectrums, the line spectra and the continuous or "white" spectra. The line spectrum is specific to the material under investigation and has specific wavelengths. The continuous spectrum is the one used in radiography and is produced by the rapid deceleration of the electrons on impact and has a broad range of wavelengths [3]. Wavelength and energy are used to characterize X-rays and are related through the equation below where E is the radiation energy, h is Plank's constant, c is the velocity of light and λ is the radiation wavelength. From inspection, higher energy will have shorter wavelengths, allowing for more penetration capability.

$$E = \frac{hc}{\lambda} = \frac{1.24}{\lambda} \tag{3.1}$$

X-ray tubes are a key component in the generation of X-rays beams for traditional X-radiography. A cathode and an anode are contained within a glass bulb under vacuum. The cathode contains a wire filament which will emit a continuous stream of electrons when heated to incandescence. The anode contains a target at which the electrons are directed. It is at this target that X-rays are produced upon impact. This target is generally made of tungsten for two reasons, first, it is a good source of high-energy X-rays and second, it has a high melting point. Most of the energy used for X-ray production (99%) is converted to heat and most of this heat conversion takes place at the target so it is necessary to have target material which can withstand high temperatures.

The electrical tension between the anode and cathode causes the acceleration of the electrons to the target, and the electron stream is focused into a beam by a cylinder or focusing cup. After impact on the target, the X-rays exit through a window made of a light element, usually beryllium, that will not absorb much of the radiation as it passes. The target is oriented at an angle to the beam of electrons in order to project the X-rays out the window. The angle at which the target is oriented reduces the effective width of the target and the X-ray beam width. This will have a large effect on the image production as a smaller effective target width produces sharper radiographs.

X-ray equipment is characterized by its potential (in volts) and current; factors which control the intensity and penetration capabilities of the radiation. Typical equipment has a range from 50 up to 320 kV; equipment designed for specialized uses may range up to 450 kV. For portable units a potential of 200 kV with intensity of 3 mA is standard [1, 2].

3.2.4 Radiation Attenuation

As X-rays and gamma rays pass through material, attenuation occurs depending on the material composition, density and thickness as well as the energy of the radiation beam. This attenuation, or loss in intensity, is what makes radiographic inspection possible. The detection of the difference in radiation intensity is recorded to produce radiographic images that allow for detailed assessment. The intensity of radiation upon exit of a material is given by

$$I_X = I_O \cdot e^{-\mu \cdot t} \tag{3.2}$$

where I_X is the emergent intensity, I_O is the initial intensity, t is the thickness of the material and μ is the linear absorption coefficient per unit thickness, a material characteristic affected by the density. The absorption coefficient is frequently reported

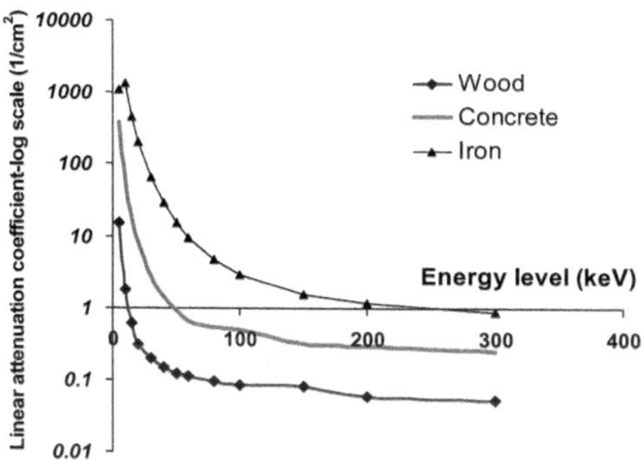

Fig. 3.2 Linear attenuation (absorption) coefficient (μ) for wood, concrete and steel [4] as a function of energy level

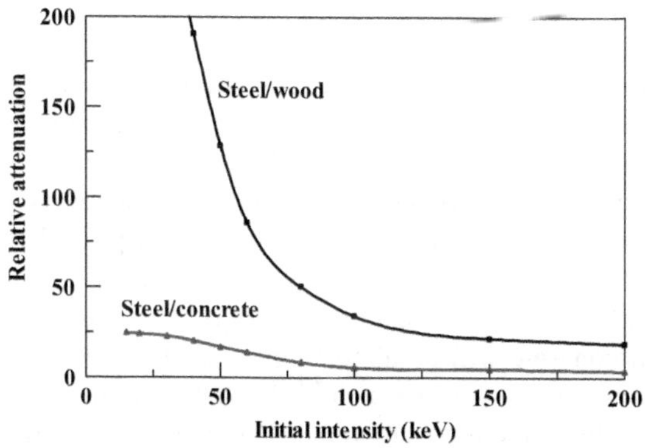

Fig. 3.3 Relative attenuation for steel, concrete and wood

as a mass absorption coefficient in which the linear absorption coefficient is divided by mass density.

Relative attenuation (composition contrast) is important along with the radiation intensity leaving the object. The relative attenuation is a function of the energy level and decreases as the energy level increases. This creates potential difficulties in materials with similar attenuation values or when a medium has a relatively high attenuation such as reinforced concrete. This problem is illustrated in Figure 3.3. The largest differential attenuation is at low energy levels but such levels may not penetrate the investigate material as it is frequently the case in reinforced concrete. Increasing the input energy levels will decrease the relative attenuation and the resulting sensitivity. The chart in Figure 3.3 is normalized assuming unit thickness. For example at 50 keV the ratio between steel and wood thickness will have to reach about 150 for the attenuation to be equal while at 200 keV level the ratio will drop to about 25.

Composition contrast is affected by the x-ray wavelength and wavelength versus mass absorption charts can be used to find the optimum wavelength. The relation-

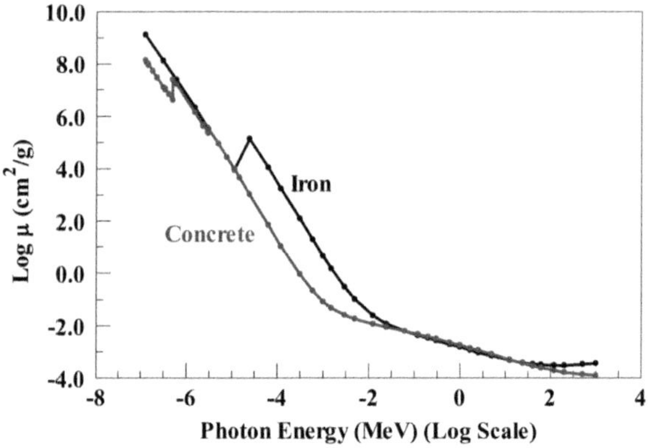

Fig. 3.4 Photon energy versus mass absorption coefficient for iron and concrete [4]

ships between mass absorption and the wavelength or mass absorption and photon energy experience discontinuities depending upon the mechanism controlling the attenuation (Thompson (coherent) scattering, photoelectric effect, Compton (incoherent) scattering, pair-production of an electron and positron, and photodisintegration). At different energy level, different mechanism may prevail and this results in discontinuities in energy-mass absorption curves (Figure 3.4).

From Figure 3.4 it is clear that at certain energy levels mass absorption coefficients of iron and concrete will be almost equal. Composition contrast will be at a minimum at x-ray energies of about 1 MeV where the mass absorption coefficients will be about the same for all materials [5]. Real-time radiography is gaining acceptance in laboratory investigations where the image is digitized via a convertor and displayed and stored at the same time [5]. Imagers with acquisitions speeds exceeding 20 frames per second (fps) are available and the acquisition rates will further increase. The acquisition rate is linked with the image quality and higher rates result in lower pixel density. The real-time radiography offers a unique opportunity to investigate transient processes both quantitatively and qualitatively [5]. While currently confined to a laboratory environment, the-situ applications are a logical extension of this technique.

Attenuation is also dependent on the radiation energies. Radiation with low energies are more readily absorbed and prone to scatter, resulting in less penetrative power. In contrast, higher energy beams will be more penetrative with less absorption and subsequent scatter.

3.2.5 Imaging

Radiographic images are produced based on the intensity of radiation exposure on an imaging plane. Images can be permanently recorded using traditional film or paper mediums, or sensitive real-time imaging mediums integrated with digital systems and software.

Film radiographs are the traditional form of capturing images. These films have an emulsion that reacts and changes when exposed to radiation. Upon development, a negative image or "shadow image" is produced where denser areas, which allow less radiation exposure, appear lighter. This form of imaging has been limited for in situ evaluation when used with the traditional high-energy radiations for safety concerns as well as the high cost of the operation; however it does provide a permanent record of the investigation and film is relatively inexpensive to purchase and process.

Radioscopy, or real time imaging, was one of the first forms of radiographic imaging. Traditionally florescent screens were used with high-energy radiation sources to produce an image based on the ensuing radiation. The screen emitted light based on the radiation it was exposed to; brightness being proportional to the intensity of the ensuing radiation, producing a positive image. This method was more portable than film radiographs and offered the advantage of real-time images that could be utilized to improve the inspection. There was however safely concerns associated with the high-energy radiation source and such a technique was unable to record the images for future analysis.

With technological advances digital radioscopy has emerged as a viable assessment option without previous drawbacks. The radiation source can be of a lower energy and detected radiation can be recorded on reusable imaging screens and processed into digital images that can be stored for future use. The digital storage of images allows for powerful image enhancing tools to be utilized which can provide more detail and allow for further information extraction.

3.2.5.1 Image Quality

High contrast and sharpness are desired and make inspection and interpretation of the radiographs easier. Geometric features relating to equipment and object positioning will affect the image sharpness. The image property known as geometric unsharpness, U_g, is given by an equation with variables S denoting the size of the focal spot within an X-ray tube, a denoting the distance from the source to the object, and b denoting the distance from the object to the imaging material.

$$U_g = S\left(\frac{b}{a}\right) \tag{3.3}$$

Sharpness and image quality will decrease if the focal spot in the radiation equipment is large and with the increase in distance between the object and recording medium. Distance between the object and the recording medium should be kept at a minimum to improve image quality and to avoid image magnification and distortion. The distance b can be increased to produce a magnification of the object to inspect small features, but this should be done carefully to keep the unsharpness at an acceptable level [1]. The contrast of images is the amount of difference seen between densities and is an important quality issue since good contrasts distinguishes member features. Contrast of radiographic images is highly dependent on the recording material and energy levels of radiation. Lower energy radiation produces higher quality contrasts, but is limited in its penetration ability and the range of densities it can produce on an image. Higher energy levels, while more penetrative, will have less contrast in images. Selection of radiation energy will most likely depend on the material investigated as well as the detail needed in the radiographs.

Image quality can be greatly affected by the selected view (the source, object, imaging plate relation). In general it is best to orient the radiation source and the image capturing material at right angles to the object surface to avoid gross distortions which make interpretation difficult. This however is not always plausible or desired based on the in situ member orientation and shape, the available access, or specific areas of interest on the object that do not lend themselves to this type of arrangement. This can cause increased difficulty in radiograph interpretations as a result of distorted size, orientation or overlapping images. It is helpful to place identifying markers on the imaging planes as a reference point to help identify the proper orientation of the image. This is true not only in cases where interpretation is suspected to be difficult, and it makes future interpretations of the images easier if the marker orientation is known or standardized. These markers should be of a material that will easily appear on the radiograph and should be placed on the outer edge of the imaging planes to avoid interfering with the area of interest.

Orientation of the radiation beam must also be carefully considered for deterioration or crack detection. It must also be noted that for crack detection, the radiation beam must be parallel to the crack or the crack must be sufficiently large to be detected.

3.2.5.2 Image Enhancement

Digital imaging systems offer the ability for image enhancement. Laptop computers can process the images onsite many times and perform numerous manipulations to improve image clarity. Manipulations would include contrast adjustments, brightness, color processing, figure orientation and magnifications. A feature useful for image interpretation reverses the gray scale so that darker areas will correspond with areas of higher density, producing a more intuitive represented image of the member. Software can also produce grid overlays and measure image features on screen. On-site imaging and enhancement also gives inspectors the advantage of viewing their work to make further images and adjustments as necessary for their needs.

3.3 Application

Radiography has been used since the 1960s for defect and deterioration detection investigations of in situ structures. Original use of high energy X-ray sources limited radiographic investigation, but the development of digital radioscopy systems has increased its use due to reduced safety concerns and cost, the ability to produce images nearly in real time with reusable imaging plates, and the capability to perform initial image assessment and manipulation on site [6]. Application on notable structures includes Thomas Jefferson's Academicals Village at the University of Virginia, Monticello, and the Narbonne house in Salem Massachusetts. Radiographic images were used to investigate the timber condition as well as verify the existence and condition of metal fasteners and hardware and answer questions of internal or hidden construction techniques [6, 7]. For this application, the interest lies in X-ray use to locate deterioration in timber members and discussion will be limited to that topic.

Attenuation is a function of the radiation energy, member thickness and, important for material condition inspection, density. Timber condition can be assessed by examining the density variations of radiographic images. Decay will appear as areas with less density resulting from the breakdown of the material.

Stages of deterioration can be identified through the examination of the radiographs. Sound wood will present a clearly defined wood structure including annular rings or grain, and optical density will be uniform. Partial decay will show loss in the wood structure, annular rings will appear but will be vague, and the optical density will vary over the material showing areas of density deviation. Decayed areas will have lost the wood structure, appearing only as an amorphous mass. Horizontal separation lines will appear resulting from the material breakdown, and decay pockets will be identifiable. Advanced decay will have the same features, but they will be more severe and extended through the member [8].

High resolution images of wood members can be produced which will show density variations between early wood and late wood. This distinction in density variation allows for grain to be visible in the images, as shown in Figure 3.5. Variations or loss in grain distinction can be used to assess timber construction as well as identify areas where deterioration or infestation has set in. Figure 3.6 shows an example of a deteriorated member; the lighter portions of the image correspond to loss in density due to decay, the grain pattern in lost in this area as well.

Disruption in grain structure can be used to locate internal features such as knots and grain deviations. Insect damage has been identified and located using radiography on timber members. Mechanical damages, such as fractures, drill holes, cuts, or naturally occurring cavities can also be identified.

Fig. 3.5 Example of sound wood with wood grain visible in X-ray image, gluelam beam

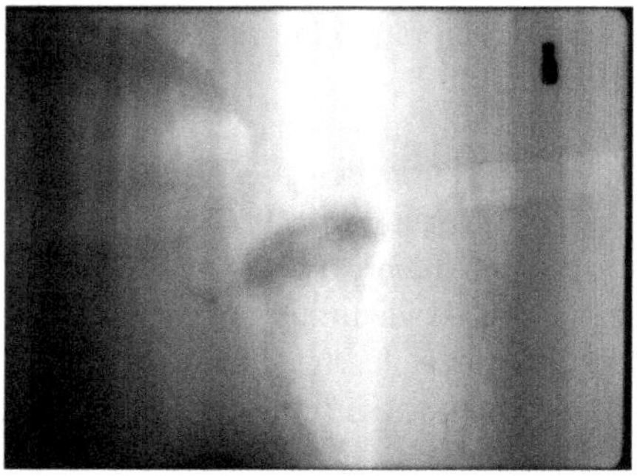

Fig. 3.6 Example of deterioration visible in X-ray image

3.4 Limitations

While radiography offers the ability to view internal characteristics of members, there are limitations and disadvantages to the technique. Radiographic investigation can commonly identify deterioration and defects in timber material; however defect depth can be hard to establish, and the extent of deterioration cannot be quantified. Radiographic images produce a two dimension representation of the inspected timber member by compressing data through the thickness into one plane. Therefore the density data and images produced represent the average density of the member through the thickness. This fundamental imaging process does not allow for information to be gathered on the depth of internal features and can make detection of cracks or defects that are oriented perpendicular to the radiation path difficult. This also makes estimates on the amount of material lost to degradation and decay difficult based on single images of the member. Sound wood structure can also be superimposed over deteriorated areas and make interpretation and quantification difficult.

More images, taken from different angles and on different faces, would help in gaining perspective on the extent of internal damage. This would raise the time and cost associated with the investigations. Research into the ability to quantify deterioration through image manipulation and radiographic data is, however, being investigated currently.

Detection of internal flaws using radiography can be limited by size and orientation of defects. Locating internal cracks requires that the crack be of adequate size, at least 2% of the member thickness [2], and must be oriented parallel to the radiation beam to be detected.

Although reduced with the development of low energy portable X-ray systems, safety can be a concern which can limit its use. Radiation is not detectable to human senses but it is very harmful to living tissue. With these characteristics, it is important to monitor radiation exposure when using radiography. This is especially true when using high energy sources needed for inspecting high density or thick members. Safety risks can be minimized by monitoring exposure and using good protective practices with radiation. Exposure is most commonly monitored using pocket dosimeters or film badges worn by personnel at investigation sites. The level of exposure can be controlled by simple means. Restrictions should be placed on the source intensity and the emission direction and all persons in the area should be kept informed as to when exposures are done as well as kept out of the immediate area of radiation investigation [2].

Limitation on the intensity or energy level of the radiation source can also limit investigation of materials. Thick members may be difficult to examine with the low energy radiation sources that are used for in situ investigations. If the member is too thick not enough X-rays will be transmitted to produce quality images, or the time required to scan with adequate results will be excessively long [9].

Member arrangement can also make positioning of the source and imaging plates very difficult and require equipment solely for access purposes, as well as more thought and time put into the set up and interpretation of the images. Problems placing image capturing materials opposite radiation sources can arise in many situations. Most equipment currently used for radiographic investigation requires access to opposing sides of a member, which for in situ testing may not be available. View choice may also be obstructed by other structural material making the desired view of the member inaccessible.

In addition to these functional issues, radiography is also a more expensive form of nondestructive testing than many other alternative methods. Portable units make members more accessible for inspection, and allow for processing the images onsite with little to no costs, but the initial costs of the equipment can be prohibitive. Limitations on source energy will also increase time needed and costs incurred with inspection of thick or very dense members.

References

1. Lang, J. and Middleton, A. *Radiography of Cultural Material*. Butterworth-Heinemann, Burlington, 1997.

2. Raj, B., Jayakumar, T., and Thavasimuthu, M. *Practical Non-Destructive Testing*, 2nd ed. Narosa Publishing House, New Delhi, pp. 77–99, 2002.
3. Bucar, V. *Nondestructive Characterization and Imaging of Wood*. Springer, Berlin, pp. 181–213, 2003.
4. Chantler, C.T., Olsen, K., Dragoset, R.A., Chang, J., Kishore, A.R., Kotochigova, S.A., and Zucker, D.S. X-Ray Form Factor, Attenuation and Scattering Tables (version 2.1). Online available: http://physics.nist.gov/ffast (accessed 22 March 2008). National Institute of Standards and Technology, Gaithersburg, MD, 2005. Originally published as Chantler, C.T., *J. Phys. Chem. Ref. Data*, Vol. 29(4), pp. 597–1048, 2000; and Chantler, C.T., *J. Phys. Chem. Ref. Data*, Vol. 24, pp. 71–643, 1995.
5. Cartz, L. *Nondestructive Testing*. ASM International, Materials Park, OH,. 229 pp., 1995.
6. Anthony, R.W. Use of portable X-ray equipment to investigate historic timber structures. In: *Proceedings of the Workshop on Historic Roof Timber Frames*, Trest Castle, Czech Republic, 2004.
7. Kasal, B. and Anthony, R. Advances in in situ evaluation of timber structures. *Progress in Structural Engineering and Materials*, Vol. 6(2), pp. 94–103, 2004.
8. Mothershead, S. Applicability of radiography to inspection of wood products. In: *Proceedings of 2nd Symposium on Nondestructive Testing of Wood*. Washington State University, National Science Foundation, Spokane, WA, pp. 307–336, 1965.
9. Davis, J. et al. Microstructure of wood using high resolution X-ray computed tomography. In: *Proceedings of 8th International Symposium on Nondestructive Testing of Wood*. Washington State University, USDA Forest Products Laboratory, Vancouver, WA, pp. 91–103, 1992.

Chapter 4
Resistance Drilling

Gretchen Lear, Bohumil Kasal and Ron Anthony

Abstract Resistance drilling is based on the correlation between cutting (drilling) resistance and material density and hardness. Small needle-like drill bit is driven into the material with a constant speed and feed rate. Energy needed for the drill to proceed is recorded as a function of the bit position. The resistance-position plot is then used to make inferences on material (voids-no voids). Appropriate drill but size (diameter) must be selected to avoid bending the drill along the late wood annual rings. Correlation between wood mechanical properties and drilling resistance is poor. Core drilling is a local method and gives localized information about the member quality at the location of the test only.

Key words: drilling, drill bit, resistance drilling log, deterioration, density variation, correlation, profile, resistance measure, cross-sectional mapping

4.1 Background

The most common damage to timber members comes from deterioration which can inflict internal damage without surface indicators until the damage is considerable and severe [1]. Resistance drilling is a semi-destructive method that inflicts minimal damage to the member surface while giving information on the internal condition of timber members. Resistance drilling has been used in many applications including tree growth and health investigations, bridge and building investigations, and in the termite and pest control industry.

B. Kasal and T. Tannert (eds.), In Situ Assessment of Structural Timber, 51–57.

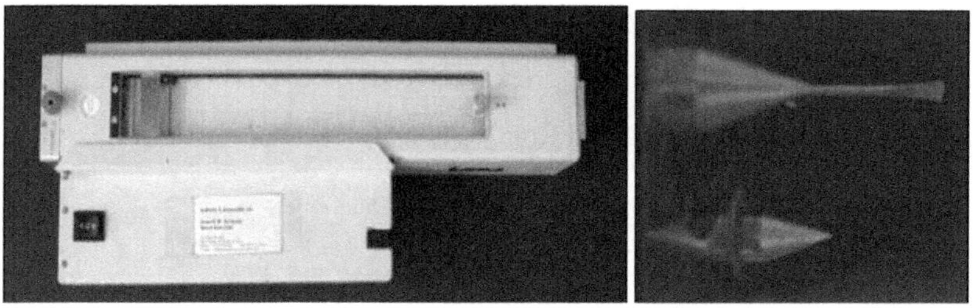

Fig. 4.1 Resistance drill (left) and drill bit (right)

4.2 Equipment

Resistance drilling offers a non-destructive means of analyzing the quality of the interior material in wood members. Resistance drills use small diameter (1.5–3.0 mm [0.6–0.12 in]) needle-like drills to bore into timber members and measures the resistance the drill bit encounters as a function of the penetrated depth. Resistance drills have electric motors and are battery operated, offering portability for field investigations. Drill bits are flexible, tungsten steel-tipped needles that will vary in length depending on manufacturer. The needle has to be replaced after 50 to 100 drillings, depending on manufacturer and use. Drill weight varies with manufacturer but is typically close to three kilograms; equipment shown in Figure 4.1.

The drill bit is advanced and rotated at a constant speed throughout the drilling. The torque required to maintain the constant cutting speed corresponds to resistance and is recorded and graphed with respect to the penetration depth [1]. Graphing of resistance data can be done with paper strips, wax paper, or recorded and stored electronically on computer. Peaks in drilling plots correspond to higher resistance or density, while dips and low points are associated with lower resistance and density; example graph presented in Figure 4.2.

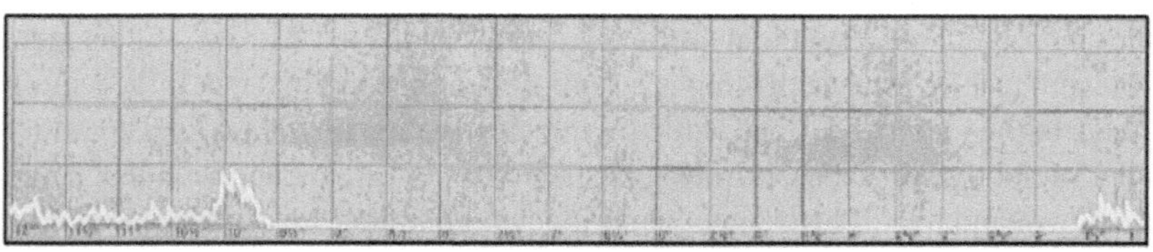

Fig. 4.2 Sample resistance drilling log

4.3 Application

Resistance drilling is used to locate and quantify deterioration in wood members. Resistance observed while drilling is proportional to density variations, or relative densities, of the timber member. Less drilling resistance requires less torque production by the motor. Areas requiring less torque are associated with reduced density such as points with interior voids, deterioration, splits or cracks [2].

Logs of multiple drillings from different directions, over the cross-section, and along the length of the member can be compiled and organized to map the condition of member. These plots can define the cross-sectional condition at a point along the member and/or the plots can be used to map the extent and penetration of deterioration along the member length.

Different levels and sources of deterioration will exhibit different drilling resistance patterns. Peaks and valleys will differentiate between areas of high and low densities. Total decay will offer no resistance and the drilling profile will appear as a zero flat line where voids are present. Deteriorated material will show some resistance but will be below that of sound wood. Sound wood will require more torque production from the drill motor and show high resistance. Sound wood will still exhibit peaks and valleys associated with early and late wood resulting from their density variations.

Insect damage will also present distinct drilling patterns. As insects, such as termites, target different parts of wood, they leave some areas intact and others void. Drill plots will show repeating spikes in resistance as the drill passes through the sound wood left behind and the voids resulting from the insect infestation [3].

High correlations between resistance drilling and gross density can be achieved. The profiles of current devices is printed with respect to drilling depth in a resolution of up to 1/100 mm and achieves correlations of even $r^2 > 0.9$ between mean resistance value and gross density of the penetrated wood [4] (see Figure 4.3).

Resistance drilling only delivers reliable results, if applied at the appropriate point, in the correct way and if interpreted regarding wood anatomical properties of the wood. Due to growth ring structure, beams of ring porous species have a higher density in the center. Conifers are mostly softer inside, thus the profiles drops should not be misinterpreted as incipient decay.

With some resistance drills the user has to select the drilling speed regarding the density of the wood. If the drilling is too slow, the profiles do not clearly reveal changes between intact wood and decay. If drilling is too fast, the needle may break. In addition, profiles made with different speeds can not be directly compared due to different ordinate scaling. Thus, selection of the optimum drilling speed is a critical point and requires experience.

High-resolution drill versions maximize their drilling speed automatically during drilling. In "soft" density wood, the needle penetrates with maximum speed, in "hard" wood it slows down. These drilling machines pull the needle back automatically if it reached maximum penetration depth, and, in addition, the obtained values are calibrated to the same scale and enables the user to compare profiles derived from different species.

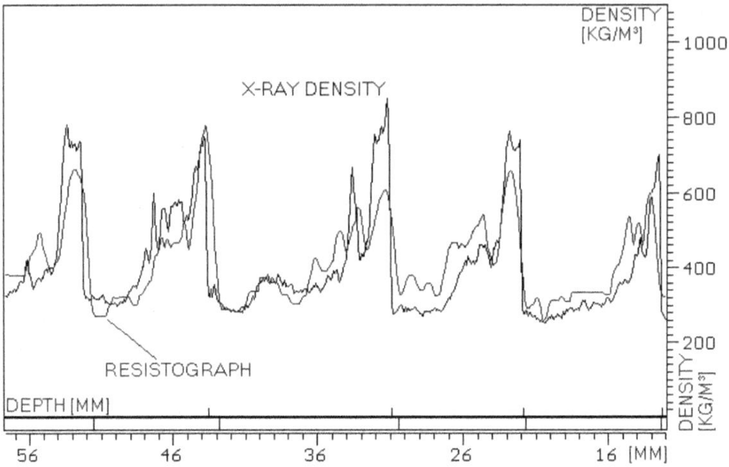

Fig. 4.3 Profiles of radial x-ray density and drill resistance of the same larch sample [4]

Figure 4.4 shows examples of drill resistance profiles; top left: drill resistance profiles from spruce where the tree rings are more or less visible depending on the penetration angle, top right: drill resistance profiles from tangential drillings in oak that look like radial drillings in conifers (thus it is important to know the species, drilling point and drilling angle to be able to interpret the profiles reliably), centre left: strong decay in historic oak beam revealed by all profiles, centre right: two or even more drillings are required to reliably assess the internal condition, bottom left: light insect damage revealed by some depressions in the profiles that can be missed by only one drilling, and bottom right: strong insect damage identified by a succession of many typical depressions.

While resistance drilling offers a sound method for detecting and measuring interior damage in timber members, it offers no quantification of mechanical properties. Studies have shown that the correlation of the drill resistance to the density of the timber member is variable, citing also weak correlation coefficients between $r^2 = 0.21$–0.69 from [2]. Even with strong correlation between drilling resistance and density values, the relationship between density and mechanical properties is not well defined.

Feio et al. [5] performed comparative tests between destructive tests and resistance drilling and used a parameter termed a Resistance Measure (RM) which allowed for the resistance drilling results to be compared to the density and strength values obtained destructively:

$$\mathrm{RM} = \frac{\int_0^h \mathrm{Area}}{h} \tag{4.1}$$

The RM parameter is the ratio between the integral of the area of the drill diagram and the height of the test specimen h. Feio et al. [5] showed that the correlation

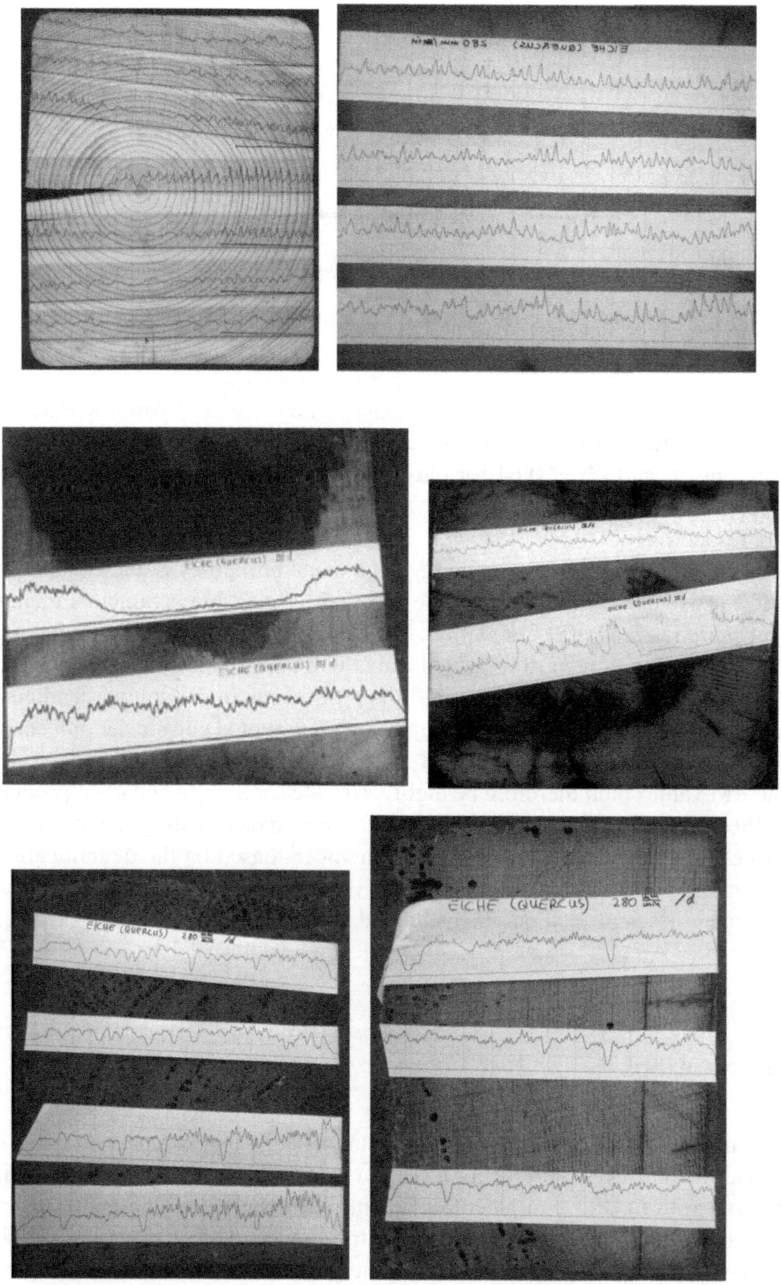

Fig. 4.4 Examples of drill resistance profiles (pictures courtesy of Frank Rinn)

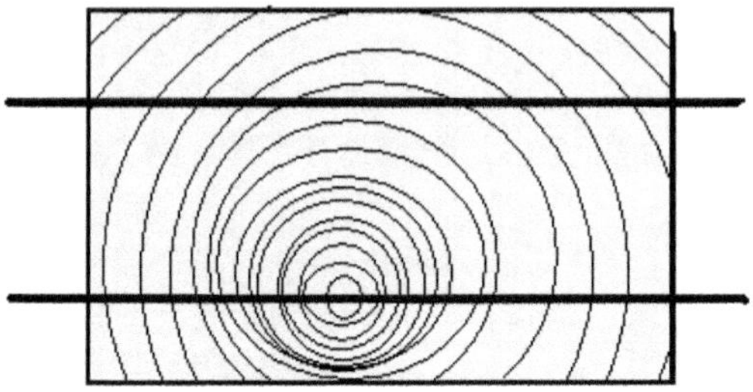

Fig. 4.5 Drill orientation effect on annular rings encountered

between the RM value and measured density to be $r^2 = 0.71$ for new timber and $r^2 = 0.68$ for old timber. RM had a correlation to modulus of elasticity of $r^2 = 0.60$ for new timber, and $r^2 = 0.64$ for old timber. Correlation of RM to longitudinal compressive strength was shown to be $r^2 = 0.59$ and $r^2 = 0.64$ for new and old timber, respectively.

The use of the RM value to estimate mechanical properties is questionable. The area of a resistance drilling plot can be affected by multiple parameters including drill bit sharpness and drill orientation. On a single member, changes in the orientation of the drill with respect to growth rings will change the calculated RM value with each drilling. Drill orientation into the pith results in drilling perpendicularly through many tightly spaced growth rings, drilling oriented outside the pith encounters less growth rings which are spaced out (see Figure 4.5).

The RM value could therefore be highly variable for a single member, producing variable estimates of the mechanical properties. Resistance drilling can, however, be used to estimate mechanical properties of members based on the quantification of deterioration and basing calculations on the remaining sound material and information on clear strength that can be estimated with other non- and semi-destructive testing.

4.4 Limitations

As with most investigative techniques, limitations exist that can prevent the use of resistance drilling. Accessibility is a common factor and includes issues such as limited space to fit drill equipment, achieving the required alignment to drill desired areas with limited space, complete obstruction from other structural members or elements, and issues of location such as height or orientation that makes the option of drilling unavailable. Equipment itself might limit accessibility based on the length of drill bit with regards to the member dimensions and drilling the full desired dimension. Drilling equipment is also relatively heavy and cumbersome to the operator and can cause fatigue after multiple drillings and extended use.

Cross-sectional mapping can be very useful for locating and identifying extent of internal damage, but it can require multiple, time consuming drillings. Multiple drillings, if any are allowed at all, may be limited by member surface décor, significance, or property owner objections, reducing mapping ability. Although the needle is thin, the drilling hole may be too big for very important historic timber elements.

The small diameter drill needle has a low stiffness making it flexible. The drilling needle can bend and follow growth rings as it penetrates the member. This can cause deviation in the drilling path and inaccuracy in drilling profiles, especially if the path deviation goes undetected. Because the results are only valid for the drilling spot, a well trained operator should select the most appropriate points where to drill and how to drill (in which direction and which angle) and how to interpret the profiles.

If the needle is damaged or was used too often, it has to be replaced. Blunt needles tend to deviate much more from the straight direction, leading to profiles that can not be evaluated properly. Only experienced operators hear such deflections during drilling by the noise caused by the high shaft friction and they can interpret the increasing profile level.

Although high quality resistance drills can achieve good correlation between mean resistance and gross density, these values do not cover knots, fiber deviations and other aspects that have to be taken into account, if strength values are needed. Thus resistance drilling can only be one part of such a job. If the task is just to find decay, even in hidden beams, properly applied resistance drilling can provide valuable information in short time – in addition to visual inspection. If strength values are requires, the best drills available on the market are required as well as a high level of education and experience. Thus, proper education of the user is a critical point before applying resistance drilling.

References

1. Brashaw, B., Vatalaro, R.J, Wacker, J.P., and Ross, R.J. Condition assessment of timber bridges: 1. Evaluation of a micro-drilling tool. Gen. Tech. Rep. FPL-GTR-159. US Department of Agriculture, Forest Service, Forest Products Laboratory, Madison, WI, 2005.
2. Kasal, B. and Anthony, R. Advances in in situ evaluation of timber structures. *Progress in Structural Engineering and Materials*, Vol. 6(2), pp. 94–103, 2004.
3. Vermeer Australia. The IML Resistograph, http://www.vermeeraustralia.com, 20 July 2005.
4. Rinn, F., Schweingruber, F.H., and Schär, E. Resistograph and X-ray density charts of wood. Comparative evaluation of drill resistance profiles and X-ray density charts of different wood species. *Holzforschung*, Vol. 50, pp. 303–311, 1996.
5. Feio, A.O., Machado, J.S., and Lourenco, P.B. Parallel to the grain behavior and NDT correlations for chestnut wood (Castanea Sativa Mill). In: *Proceedings Conservation of Historic Wooden Structures*, Florence, Italy, pp. 294–303, 2005.

Chapter 5
Core-Drilling

Bohumil Kasal

Abstract Core drilling is a semi-destructive method where cores of small diameter of about 5 mm are extracted from members and tested in a compression device. Small core tests are correlated with the standard small-clear specimen tests to evaluate compressive strength and modulus of elasticity. Cores are smaller than natural defects such as knots and can be plugged after testing. This is a local method that gives localized information about the mechanical properties of the clear wood. The method requires special testing fixture and testing equipment.

Key words: cores, core specimen, mechanical properties, compressive strength, modulus of elasticity, density, hollow bit, load-deformation plot, correlation, number of samples

5.1 Background

The quantitative core drilling method was pioneered in the early 1980s by Dr.-Ing. Jaroslav Lexa at the State Forest Products Research Laboratory in Bratislava, Slovakia and furher improved by Kasal [1]. Core samples are most commonly circular specimens tested in compression to establish the compressive strength of the material. Core samples provide a local property value that can then be used to make inferences on the member's overall strength.

Equipment includes manual or electrically driven drills for core extraction, as well as load cells and fixtures for testing the cores. Core samples will vary in size depending on equipment, but the premise behind this semi-destructive testing technique is that the extraction of cores will leave holes that are smaller than most knots found in timber members and will not compromise the strength. Voids left by drilling should be plugged to prevent moisture and insect penetration, to reduce the likelihood of introducing decay at the coring location, to restore some of the

B. Kasal and T. Tannert (eds.), In Situ Assessment of Structural Timber, 59–66.

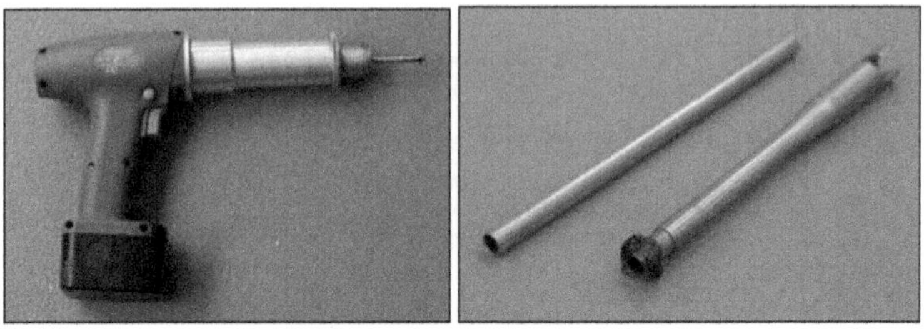

Fig. 5.1 Mechanical core drill for extracting samples for destructive testing (left) and Mechanical core drill bit (right)

minimal compressive strength lost and to help to preserve the appearance of the member [1, 2].

Core samples can provide multiple indicators of the member's general condition and wood anatomy, as well as be destructively tested to extract mechanical property information used to make inferences about the whole member.

5.2 Equipment

Incremental borers were originally used for measuring the growth and investigating the health of trees. Borers can be manual or electrical, but both utilize a hollow bit made of steel to produce small radial cores. Manual corers are auger-like tools operated with a brace-and-bit motion. To prevent breakage of drill bits, starter holes can be made with a punch hammer or an electric drill. Drilling can also be made easier by applying soap or wax to the drill bit.

Hollow drill bits can be specially fabricated to fit on standard power drills [3]. This can greatly reduce the time required for core sampling. Speed should be monitored so that drill bits and cores are not damaged, especially in dense woods or if metal fasteners could be present. Drill bits, either manual or power driven, should be kept sharp and clean to maintain quality core sampling. Dull or dirty bits can cause core samples to appear damaged or decayed as well as cause cores to jam within the bit. After taking samples, the cores should be stored and transported in safe containers. Containers should be appropriate for the samples and provide adequate protection and air tightness if required. Containers should also be labeled with member, date, location, and other pertinent information [3].

A core drill was developed to extract quality timber core samples for compressive strength, parallel-to-fiber testing [4]. Core samples are taken from clear sections of the member and are drilled to produce a specimen with annular rings oriented perpendicular to the longitudinal axis of the specimen. Proper drilling and annular ring orientation is important in order to obtain samples that can be tested along the fibers.

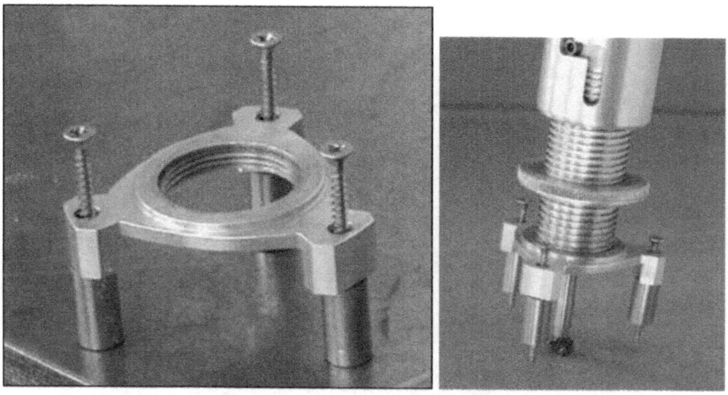

Fig. 5.2 Threaded fixture to attach core drill to timber member (left) and mechanical feed to maintain constant cutting speed while drilling (right)

Friction forces, which can become large during drilling and make quality core extraction difficult, were minimized with the design of a drill bit which has an increasing diameter along the length from tip to shank. To prevent lateral motion of the drill during extraction, the drill is attached to the member with a special fixture, and a mechanical feed is used to maintain a constant drilling speed (shown in Figure 5.2 [1,2]).

Resulting core specimens have a diameter of 4.8 mm (0.19 in) and the hole left in the member is 10 mm (0.39 in) in diameter, corresponding to the largest diameter of the drill bit. The length of the cores must be at least 20 mm [0.79 in] in order to minimize the bias of annular rings due to the variation in density between early and late wood. 20 mm (0.79 in) corresponds to the European standards length requirement for compression testing [1].

Fixtures for the compression testing of the cores along the fibers were also designed and fabricated in order to apply the compressive load parallel to the timber fibers as well as distribute the loading along the length of the core (see Figure 5.3). The specimen is placed into the cylindrical space between the compressive jaws with the fibers oriented parallel to the loading; loading applied by a load cell. A gap is left between the two compressive jaws in order to allow the specimen to deform. Two miniature linear variable differential transducers are used to monitor the displacement between the test fixtures and measure the deformation of the core.

5.3 Application

Small diameter cores are generally extracted from members and tested in compression, but a variety of properties can be established including density, moisture content, modulus of elasticity and other strength properties. Cores are routinely used for species identification through microscopic investigation, dendrochronology, micro-

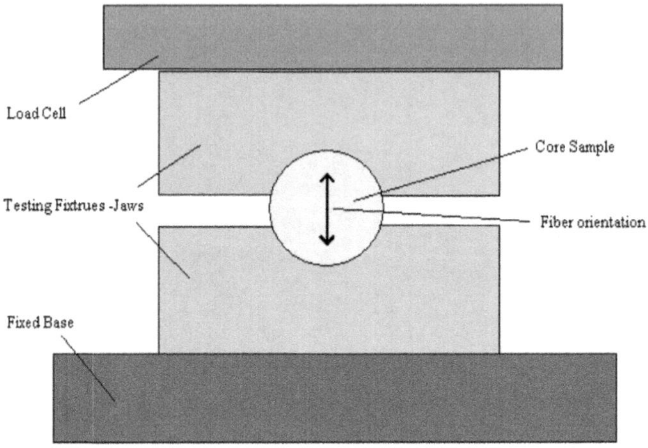

Fig. 5.3 Schematic of fixture used for compressive testing of core samples

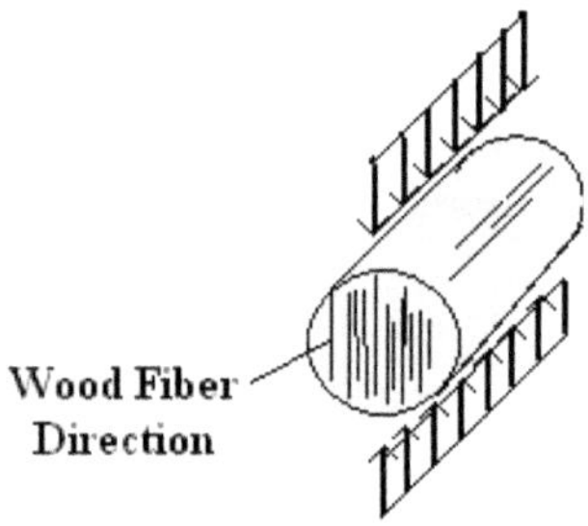

Fig. 5.4 Load orientation for compression testing of core samples

scopic inspections for signs of early decay, as well as visual examination and measurements of preservative treatment penetration and retention in the member [5]. For this application, core samples are use to establish mechanical properties and discussion will be limited to the compressive testing of cores.

Core samples are extracted from clear sections of a member and destructively tested to gain information on the mechanical properties of the member. Core samples of materials such as concrete, mortar or masonry which are considered isotropic and homogeneous can be tested along the longitudinal axis of a core sample for compressive testing. This cannot be done with timber cores since wood is an anisotropic material whose properties are directionally dependent. Strength along the wood fibers is most critical because it directly controls parameters such as bending, tensile, and compressive capacities. To achieve testing that induces load along the wood fibers, the cores must be oriented so the load is applied perpendicular to the longitudinal axis of the core (see Figure 5.4).

Proper alignment of the core sample in the testing fixtures is crucial for accurate compressive strength estimates and modulus calculations. Timber has its greatest

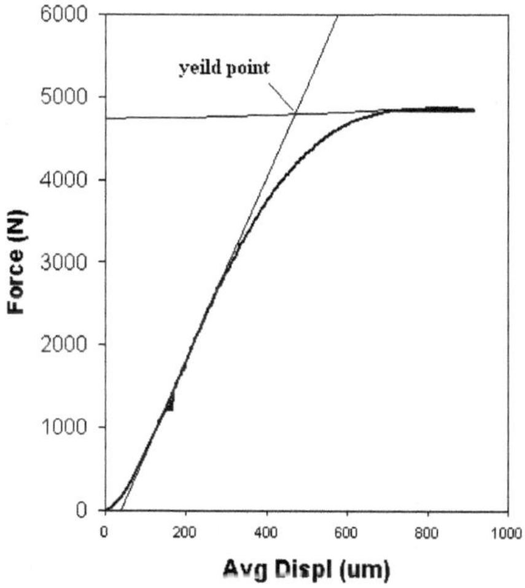

Fig. 5.5 Example load-deformation plot for core compression test parallel to fiber

strength along the fibers; misalignment of the core sample in the fixtures creating an angle between the fibers and loading will cause reduced estimates of the mechanical properties. Even slight misalignments will affect the data gained from the testing and result in conservative property approximations [1].

During testing, the compressive force and the deformation of the core are monitored and recorded to produce a load-deformation curve; sample plot shown in Figure 5.5. The modulus of elasticity cannot be directly calculated from the load-deformation plot because of a multiaxial stress field created due to surface restraints, Poisson's ratio and the geometry of the specimen. The slope of the load-deformation plot maps directly to the modulus of elasticity, a relationship presented in [1].

Calculation of specimen apparent compressive strength utilizes the following equation:

$$f_c = \frac{F_{\max}}{l \times d_c} \tag{5.1}$$

This equation reflects fundamental mechanical principles: stress is equal to force divided by area. f_c is the apparent compressive strength of the core, $F_{\max}$ is the failure load, and l and d_c are the length and diameter of the core, respectively; f_c is considered to be the apparent compressive strength because the pressure distribution over the core sample cannot be made uniform during testing as well as end effects that cannot be eliminated [1].

The failure load, $F_{\max}$, is taken as the yield point of the load-deformation curve. Due to surface restraints, Poisson's effect and the geometry of the core, a distinct yield point does not exist. The yield point is therefore defined as the intersection of the lines extended from the two quasi-linear portions of the plot [1], illustrated in Figure 5.5.

Kasal et al. [4] performed standard compressive and tensile testing in accordance to the American Society of Test Materials (ASTM) specifications [6] on specimens and compared the results to data found with core sampling and testing of the same member. A strong relationships between the core and ASTM specimen compressive strength and modulus was shown with correlation coefficients of $r^2 = 0.89$ and $r^2 = 0.76$, respectively.

Theoretically, the correlation between cores and ASTM samples should be 1.0; however, errors in correlation can be attributed to the destructive nature of the testing as well as the variation in the material along the member. The destructive nature of the testing did not allow for the samples to be used for both ASTM and core testing. Samples were taken from the same vicinity of the member however, variation between samples still existed due to the natural variability of timber [1]. Deviation in loading angle with respect to the grain during core testing would also incur error in correlation. As stated previously, wood fiber alignment can be difficult and strength estimates are sensitive to any deviations, causing reduction in predicted strength values.

Correlation between core compressive strength and the tensile strength of ASTM samples was not particularly strong. Kasal [1] cut core samples in the vicinity of the necked portions of the ASTM tension samples after testing and found a correlation coefficient of only $r^2 = 0.67$. The low correlation can be related to the ASTM sample itself. The strength of the ASTM small clear sample, with a specified cross section of 4.8×9.5 mm (0.19×0.37 in), will be affected by the amount of early and late wood present in the sample. The sample cross section may not be large enough to contain an adequate amount of annular rings to remove the bias of the early and late wood effect, especially for species with wide annular rings [1]. Kasal also noted the precarious nature of relating the compressive strength to tensile strength due to the difference in loading methods and failure modes which can lead to error in correlation.

Random sampling of the member must also be considered to gain an accurate representation of the member strength. This can be done through dividing the member into sections, assigning numbers to each segment, and then using a random number generator table to select areas for sampling.

5.4 Limitations

Limitations of the core drilling techniques for establishing material properties include the local characteristic of the data gained and accessibility issues. As with any technique that utilizes small specimens, the core samples taken from a member give information about the specific location from which it was taken. Timber can have significant variability between and within species as well as along a single member. In order to make global estimations of a member's mechanical properties based on the core drilling technique, multiple samples must be taken along the length of the structural element. Multiple samples also allow for the effects of individual irregularities in cores to be averaged out during the testing.

Large numbers of samples may also be required to establish a certain degree of reliability. Increases in drilling and sample extraction will increase time requirements on site and during testing which can increase the expense of the investigation. In addition, increased drilling, if drilling is allowed at all, may not be permitted in historically significant buildings or structures of importance for reasons of aesthetics or disruption to member fabric. Limited sampling can reduce the reliability of property estimations and reduce the usefulness of the technique.

Access to in situ members can deter the use of the core drilling procedure. Fixtures are used to secure the drill to the member and prevent lateral movement and regulate speed during drilling. Member location could make the attachment of the securing equipment difficult, as well as create a complex or awkward position for drilling. Members in small enclosed areas may also prohibit the use of the drill if space will not allow for drilling perpendicular to the surface. The drill attachment also requires a flat surface on the member and a large enough area for attachment. The dressing, or cut, of the member may not provide a flat surface with enough area to drill in a position which is perpendicular to the fibers, such as near the edge of members where space to attach the drill is limited or when drilling at an angle to the surface is required.

Cores should be drilled in the radial direction in order to produce samples, which will be useful for testing the compressive strength along the fibers. This will require surface inspection of the member and its grain to locate the appropriate position for drilling. This could be difficult if the grain is not pronounced, well defined, or a surface treatment such as paint obscures the grain pattern.

Core testing is intended to give data on the clear wood compressive strength of the member; therefore a clear sample must be taken and used with this technique. Core samples must also be taken away from defects or discontinuities in the member. This can be difficult if clear areas do not correspond to the needed fiber direction, clear areas are inaccessible with equipment, the surface is deteriorated, or clear areas are obscured by member finishes.

References

1. Kasal, B. Semi-destructive method for in-situ evaluation of compressive strength of wood structural members. *Forest Products Journal*, Vol. 53(11/12), pp. 55–58, 2003.
2. Kasal, B. and Anthony, R. Advances in in situ evaluation of timber structures. *Progress in Structural Engineering and Materials*, Vol. 6(2), pp. 94–103, 2004.
3. Maeglin, R. Increment cores: How to collect, handle and use them. General Technical Report FPL 25. US Department of Agriculture, Forest Service, Forest Products Laboratory, Madison, WI, 1979.
4. Kasal, B., Drdacky, M., and Jirovsky, I. Semi-destructive methods for evaluation of timber structures. In: *Structural Studies, Repairs and Maintenance of Heritage Architecture VIII*, C.A. Brebia (Ed.), Advances in Architecture. WIT Press, Southampton, pp. 835–842, 2003.

5. Ross, R.J., Brashaw, B.K., Wang, X., White, R.H., and Pellerin, R.F. *Wood and Timber: Condition Assessment Manual*. Forest Products Society, Madison, WI, 73 pp., 2004.
6. ASTM D 143-94e1, Standard Test Methods for Small Clear Specimens of Timber, ASTM International, 2000.

Chapter 6
Shear Test of Glue Lines

Thomas Tannert

Abstract The strength of glue lines in glued-laminated timber elements can be derived by shear tests on circular core samples. Such core samples provide a local value and are used to make inferences on the member strength. Since the quality of the glue line can vary significantly within and between members, multiple samples must be taken to get reliable global estimations of a member's properties. In addition to the shear strength, the percentage wood failure, a critical index to determine the quality of the bond line, has to be determined after testing by visual examination of the failure surface. This chapter describes the equipment needed to extract core samples, the testing procedure and the evaluation of the results. Furthermore, the application of the technique and its limitation are discussed.

Key words: core, glue lines, glued-laminated timber, percentage wood failure (PWF), shear core samples, shear test apparatus

6.1 Background

Shear tests on circular core samples are used to derive the shear strength of glue lines in glued-laminated timber elements [1,2]. Samples (Figure 6.1) provide a local property value of the glue line strength that can be used to make inferences on the member strength.

6.2 Equipment

Equipment includes electrical drills for sample extraction, as well as load cells and fixtures for testing. Shear core samples are approximately 35 mm (1.4 in) in diameter. Voids left by drilling should be plugged to prevent moisture and insect penet-

B. Kasal and T. Tannert (eds.), In Situ Assessment of Structural Timber, 67–73.

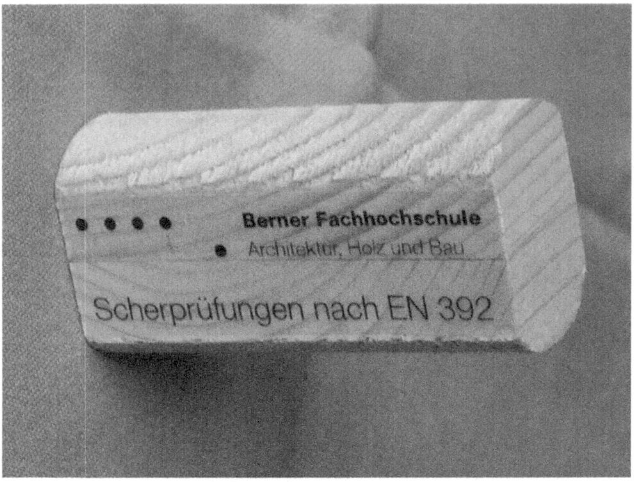

Fig. 6.1 Shear core specimen

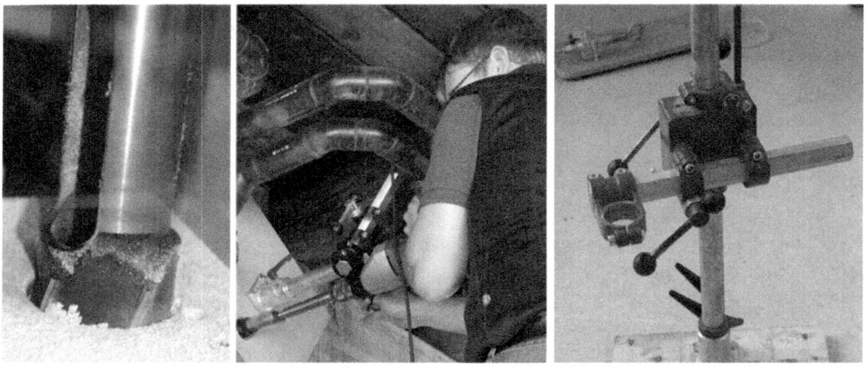

Fig. 6.2 Hollow drill bit (left), drill used for core extraction (centre), supporting fixture (right)

ration, reducing the likelihood of decay at the coring location, and help to preserve the appearance of the member.

Core drills are usually electrical with a hollow drill bit (Figure 6.2 left) made of steel and specially fabricated to fit on standard power drills (Figure 6.2 centre). The penetration speed should be monitored so that drill bits and cores are not damaged. Drill bits should be kept sharp and clean to maintain quality core sampling as dull bits can cause samples with poor surface quality. Samples should be labeled with member, date, location, and other pertinent information and be stored and transported in containers that provide adequate protection.

Samples are taken from timber components with glue lines and need to be representative for the structural member which they are taken from. Random sampling of the members can also be considered in order to get representative information. The glue line needs to be in the centre of the specimen and perpendicular to the drill

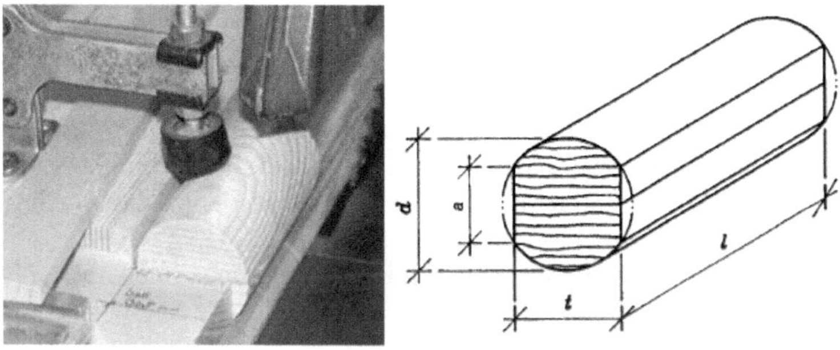

Fig. 6.3 Fixture to cut specimens (left) to final dimensions (right)

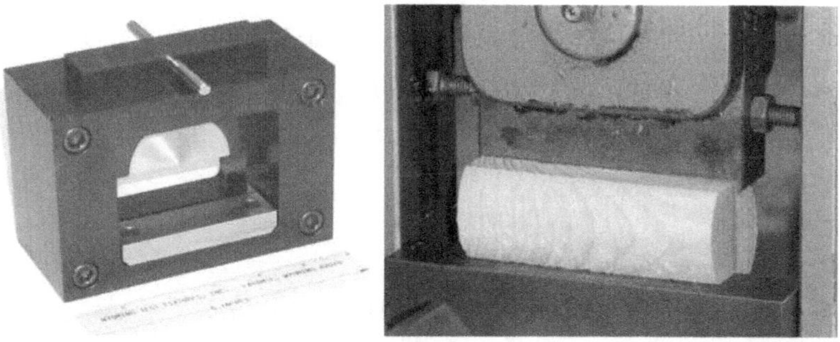

Fig. 6.4 Test fixture by Wyoming Test Fixtures (left) and specimen in test apparatus (right)

axis. Proper drilling and orientation is important; a supporting fixture (Figure 6.2 right) can be used to guarantee a correct drilling position and prevent lateral motion of the drill during extraction.

Resulting core specimens have a diameter of 35 mm (1.4 in) and the hole left in the member is approximately 45 mm (1.8 in) in diameter. The core length must be at least 70 mm (2.9 in) corresponding to the length requirement by European standards [1]. The specimens need to be flattened on two sides (Figure 6.3 left) before testing in order to be placed into the testing fixture. The final dimensions (Figure 6.3 right) are l = 70–80 mm (2.9–3.3 in), $a \approx 23$ mm (0.9 in), and $t \approx 26$ mm (1.1 in) [1]. The exact dimensions of the shear plane need to be measured.

6.3 Application

The samples have to be conditioned according to ISO 554 at a climate of 20 ± 2°C, 65 ± 5% relative humidity. After reaching constant moisture content, the specimens are placed into the space between the jaws of a shear test apparatus (Figure 6.4 left) with the glue line oriented parallel to the loading direction (Figure 6.4 right).

The shear test apparatus has to be installed in a test machine, where the moving part must be able to freely rotate. The specimens are tested with constant loading rate. Proper alignment of the specimen in the fixtures is crucial for accurate strength estimates. Misalignment of the sample creating an angle between the fibers and loading will cause wrong estimates. The allowable deviation from the glue line to the shear plane is 1 mm. In order to be able to better recognize the glue line it can be high-lighted with a pencil.

During testing, the shear force is monitored and the maximum value (F_{max}) is recorded to calculate the shear strength (f_S) according to the following equation:

$$f_S = \frac{F_{max}}{t \times l} \tag{6.1}$$

A correction factor should be applied for specimens that are wider than 50 mm [1]:

$$f_S = k \times \frac{F_{max}}{t \times l} \tag{6.2}$$

with $k = 0.78 + 0.0044 \times t$.

6.3.1 Determination of Percentage Wood Failure

In addition to the shear strength, the percentage wood failure (PWF) has to be determined after testing. PWF is a critical index to determine the quality of a bond and is usually measured by trained personnel by visual examination of the failure surface. In specimens with colorless glue, the surfaces are treated with chemicals (first hydrochloric acid, then by phloroglucinol) to better distinguish regions of wood failure from regions of glue-line failure [3]. The areas that turn red indicate fiber failure (dark area in Figure 6.5), as the lignin changes color. To determine PWF, usually two people independently evaluate the specimens. If their results deviate by more than a certain percentage, they consult a third person. Depending on PWF, different requirements on the strength of glue lines exist [4, 5]. A higher percentage of glue failure leads to higher requirements on the strength.

The samples are tested in shear, but density and moisture content can also be determined, furthermore samples can also be used for species and adhesive type identification by performing microscopic and chemical investigation.

6.4 Limitations

Material properties derived by using the core drilling technique are only local information. As with any technique that utilizes small specimens, the samples taken from a member give information about the specific location from which they were

Fig. 6.5 Determination of percentage of wood failure

taken. The quality of the glue line can vary significantly within and between members. To get global estimations of a member's mechanical properties, multiple samples must to be taken which account for the effects of individual irregularities; large numbers of samples may be required for a certain degree of reliability. But increases in sample extraction will then increase time requirements on site and during testing which can lead to higher expenses of the investigation.

Access to in situ members can deter the use of the procedure. Fixtures are used to secure the drill to the member and prevent lateral movement but member location could make the attachment of the securing equipment difficult, as well as create a complex or awkward position for drilling. Samples should be drilled in the most critical locations close to bearings or near the apex of curved member; access to these locations is usually very difficult.

The withdrawal of drill cores is often unnecessary [6], as trained experts can recognize by the tear pictures whether the wood or the glue line is damaged and with the examination of drill cores, only the quality of the wood is examined. The glue lines almost always, apart from few exceptions, show sufficient load-carrying capacity [6]. If doubts about the load-carrying capacity of the glue-line exist, the shear tests offer a possibility of comparing the experimentally derived strength to a reference strength.

Accurate drilling and testing is required but due to difficulties in reaching the location, this is not always guaranteed. Poor alignment of the glue-line with the specimen axis (Figure 6.6 left) or poor alignment of the specimen inside the test fixture (Figure 6.6 right) will cause irregular and unreliable results.

The shear stresses in typical shear test fixtures are not evenly distributed and no pure condition of shear stress can be created, rather a combination of shear and normal compressive stresses. The results are further influenced by the test set up and even the person carrying out the test. Comparing results of different test institutes or of different operators is difficult [7].

Fig. 6.6 Poor alignment of the glue-line (left) and of the specimen inside the test fixture (right)

PWF is conventionally measured by visual examination. As a result, the measurements are subjective, making it difficult to compare test results of different studies. It was reported that the PWF varied from 20 to 100% for spruce shear blocks of a red adhesive among 12 well trained researchers from 9 laboratories in a round robin study [8]. As an alternative, image analysis technique can be used to calculate the PWF [9]. An algorithm based on the image bi-modality concept to give reasonable, consistent and accurate measurement in calculation of the PWF was developed and proved to be more representative for the average of the visually estimated values.

Another method for quantifying PWF of adhesively bonded specimens incorporates a laser displacement gage with an automated two-axis positioning system that functions as a highly sensitive profilometer [10]. The specimen is continuously scanned across its width to obtain a surface failure profile. This digital profile can then be reconstructed and analyzed with appropriate software. The device is very sensitive to most observed failures, particularly wood failure in deep layers; however, wood failures close to the glue line can be problematic.

References

1. EN 392. Glued laminated timber. Shear test of glue lines. CEN, Brussels, 1995.
2. ASTM D 905-03. Standard Test Method for Strength Properties of Adhesive Bonds in Shear by Compression Loading. ASTM International, 2003.
3. Künniger, T. Automatische Bestimmung des prozentualen Faserbruchanteils bei der industriellen Klebfestigkeitsprüfung. FFWH ProjeCT 2006.05, 2007.
4. EN 386. Glued laminated timber – Performance requirements and minimum production requirements. CEN, Brussels, 2001.
5. ASTM D 5266. Standard Practice for Estimating the Percentage of Wood Failure in Adhesive Bonded Joints. ASTM International, 1999.
6. Brüninghoff, H. Reinforcement/rehabilitation of glulam structures. Int. Holzbauforum Garmisch-Partenkirchen. http://www.forumholzbau.com/, 2007.
7. Steiger, R., Risi, W., and Gehri E. Quality control of glulam: shear tests of glue lines. In: *Proceedings of the 40th Meeting of CIB-W18*, Bled, Slovenia, Paper 40-12-7, 2007.
8. CSA/A370/SC05.3. Wood Failure Round Robin Report. Canadian Standards Association, 2002.
9. Yang, Y., Gong, M., and Chui, Y.H. A new image analysis algorithm for calculating percentage wood failure. *Holzforschung*, Vol. 62, pp. 248–251, 2008.

10. Scott, C.T., Hernandez, R., Frihart, C., Gleisner, R., and Tice, T. Method for quantifying percentage wood failure in block-shear specimens by a laser scanning profilometer. *Journal of ASTM International*, Vol. 2(8), pp. 1–10, 2005.

Chapter 7
Tension Micro-Specimens

Bohumil Kasal

Abstract The principle of this method is extracting triangular specimens (about 5 mm equal-sides triangle) along the length of the member. Small-kerf circular saw with a fixture attached to the surface of the member is used to extract the specimens. Specimens are then glued to the test blocks and tested in tension. The method gives direct values of tensile modulus of elasticity and strength for clear wood. The values are local and pertain to the tested area and surface.

Key words: specimen, triangular specimen, tension, tensile strength, tensile modulus, saw, grips, grain

7.1 Background

This method was developed by Kasal [1, 2] and is based on direct measurement of tension properties along fibers. Bending strength evaluation is an important aspect of in situ evaluation of timber members as it is one of the predominant modes of loading, but estimations of the bending properties in situ can present a challenge. With information on the member's tensile properties, bending strength estimates can be made; tensile strength has been related to bending strength and is considered to be approximately equal. Tensile properties have a poor correlation with compression properties therefore tensile strength can not be estimated using information attained with the core drilling technique [1, 3]. Specimens can however be extracted to evaluate the tensile properties of in situ members with the technique described in the following sections.

B. Kasal and T. Tannert (eds.), In Situ Assessment of Structural Timber, 75–80.

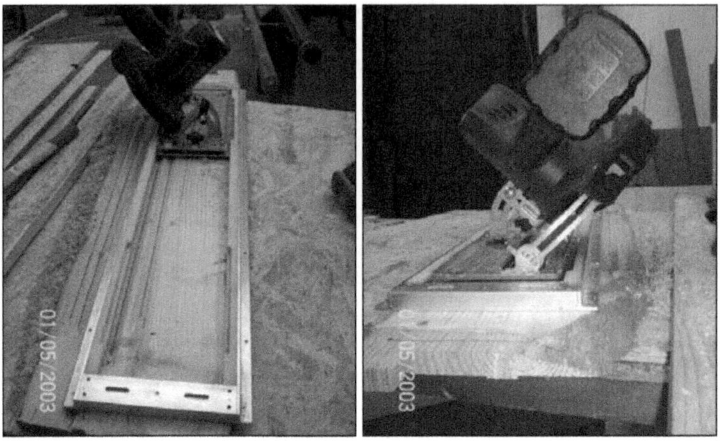

Fig. 7.1 Tension micro-specimen equipment, kerf saw and guide track

7.2 Equipment

Tension micro-specimens are made with two cuts of a thin-kerf saw oriented at 45° to the surface of the member, creating a triangular specimen. The depth of the cut, and therefore the triangle side width can be adjusted from 3 to 8 mm (0.12 to 0.31 in) to produce a sample that can range in area from 4.5 to 32 mm^2 (0.007 to 0.05 in^2). During sampling, however, cutting depths are adjusted to produce specimens with areas of approximately 8 mm^2 (0.012 in^2) so that results are directly comparable to ASTM tension specimens [1, 4]. A guide track is mounted on the member for the cut of the specimens and must be aligned to extract samples along the fibers of the member. The guide steers the saw in a straight, consistent path for the two cuts needed to produce a uniform sample and must be mounted on the member in a manner such that the path of the saw will not intersect areas with visible defects such as knots, checks or deterioration. Equipment and sample specimens are shown in Figure 7.1.

After extraction, the samples are mounted with epoxy on wooden blocks to provide a location for gripping during tensile testing (see Figure 7.2). The mounting reduces the possible influence on findings due to the local end affects of gripping the samples [1,2]. Specimens should be planed or sanded to reduce the cross section at the mid point of the samples to ensure failure at that location. Specially designed grips are used in the tension testing of the samples as well as a displacement transducer to measure the displacement during loading (see Figure 7.3).

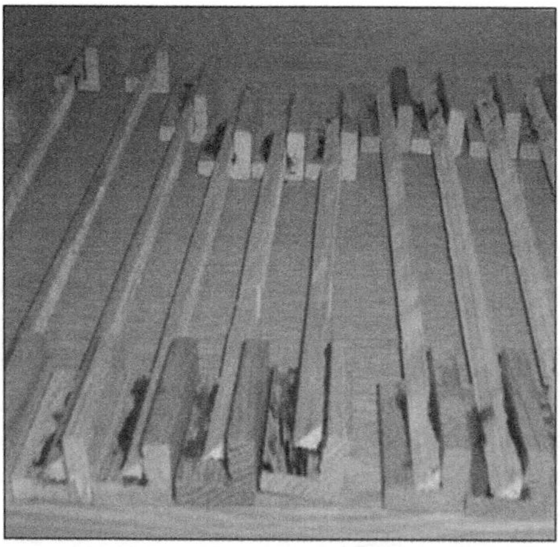

Fig. 7.2 Tension micro-specimens (mounted in grips for testing)

Fig. 7.3 Tension micro-specimens with displacement transducer attached for testing (left) and grip used during testing of tension (right)

7.3 Application

After sample extraction and preparation through sizing, sanding and mounting for experimentation, as discussed above and shown in Figure 7.2, the samples are tested under tensile loading to find the ultimate tensile strength along the grain as well as the modulus of elasticity. A schematic of the testing set up is shown in Figure 7.4.

The maximum tensile load for each specimen is the load at failure, and the allowable tensile stress is calculated by

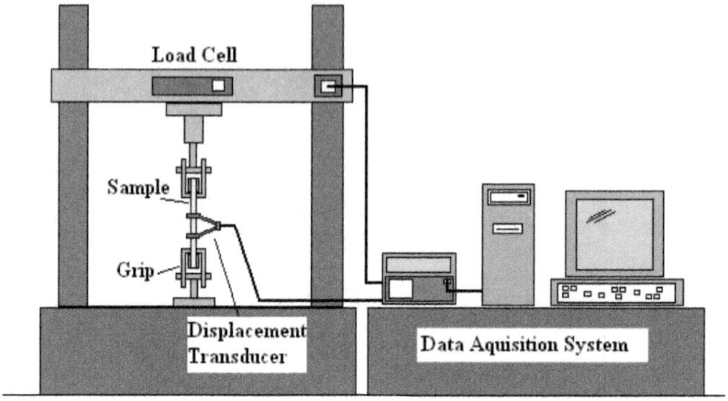

Fig. 7.4 Tension micro-specimen testing set up

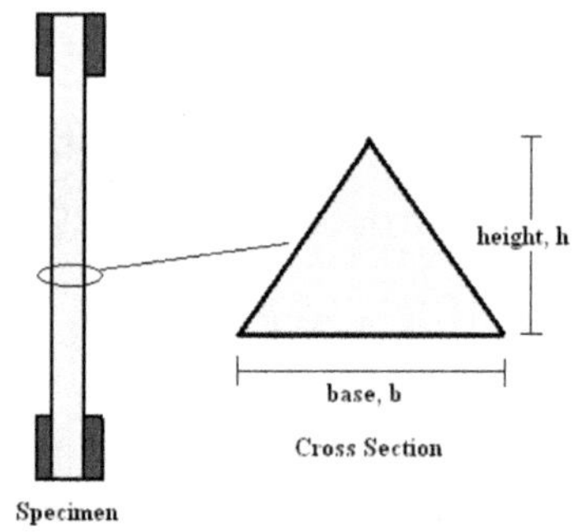

Fig. 7.5 Tensile micro-specimen cross section dimension

$$f_t = \frac{F_{\max}}{\frac{1}{2}bh} \tag{7.1}$$

where f_t is the tensile strength, $F_{\max}$ is the failure load, and b and h are the base and height of the tensile specimen, respectively (see Figure 7.5).

From recorded data, the stress-strain curve is plotted and the MOE along the timber fibers is determined as the slope of the linear trend line that fits the data (see Figure 7.6). The experimental and equipment design is such that the cross sectional area of the tension micro-specimens are comparable to the cross sectional area of the standard ASTM tension specimens for small clear wood, therefore no correlation is needed for comparison to the standard tests [1, 2].

7.4 Limitations

Tension micro-specimens are very sensitive to grain deviation. Sampling is intended to be directly along the grain of the member to achieve accurate values for

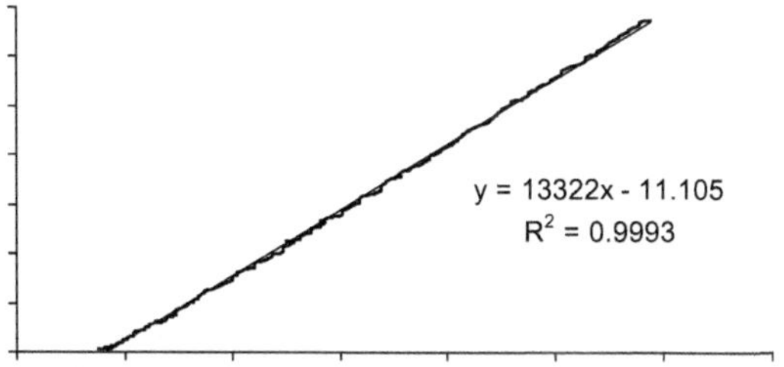

Fig. 7.6 Sample stress-strain plot of a tension micro-specimen test

tensile strength along the grain. Maintaining sampling along the grain is difficult and caution must be taken to extract quality samples. The mounting of the track guide to ensure a cut along the timber fibers is essential in extracting a quality tension sample. Specimen behavior is sensitive to grain direction and any deviation from the fiber direction will cause a reduction in the apparent strength predictions. In contrast, the modulus of elasticity is not affected by the deviation from the grain in the specimens. Testing of samples remains within the elastic limit and therefore relatively accurate measurements of the modulus of elasticity are obtained.

Cutting technique and imprecise equipment can affect the sample quality and reduce the reliability of the testing technique. More precise and accurate tension sampling equipment would improve the technique by producing more uniform, quality samples. At present, however, sample quality is a major drawback of the technique and can cause variability in test data and an inaccurate representation of member properties.

The size of the tensile specimen may also lead to some variability. The small cross-section of the specimen may not provide for the inclusion of enough growth rings to eliminate the bias of the early and late wood densities, a factor which could affect the resulting stress calculated from specimen to specimen. This source of variability could be reduced if the size of the specimen were increased, however this would increase the destructive nature of the method and perhaps limit its use in situ. This is an issue with ASTM tension specimens as well since the ASTM specimens have a comparable cross section to the tension micro-specimens.

As with all techniques that use small size specimens, the data obtained through tension micro-specimen sampling and testing is limited due to its local characteristics. Wood's natural variation within and along a member can cause deviations in material properties of samples depending on the location of extraction. To overcome the variability, relatively large numbers of samples may need to be taken in order to obtain reliable values that are representative of the member's overall tensile strength.

Selection of the sampling sites must be carefully planned to ensure a randomized sampling [1]. Restrictions on the number and location of samples that can be taken from members due to historical significance, decorative finishes or other preservation concerns can limit the use of the technique and diminish the reliability and accurate representation of the member properties.

References

1. ASTM D245-00, Standard Practice for Establishing Structural Grades and Related Allowable Properties for Visually Graded Lumber. ASTM International, 2000.
2. Kasal, B. and Anthony, R. Advances in in situ evaluation of timber structures. *Progress in Structural Engineering and Materials*, Vol. 6(2), pp. 94–103, 2004.
3. Kasal, B., Drdacky, M., and Jirovsky, I. Semi-destructive methods for evaluation of timber structures. In: *Structural Studies, Repairs and Maintenance of Heritage Architecture VIII*, C.A. Brebia (Ed.), Advances in Architecture, WIT Press, Southampton, pp. 835–842, 2003.
4. ASTM D 143-94e1, Standard Test Methods for Small Clear Specimens of Timber. ASTM International, 2000.

Chapter 8
Screw Resistance

Nobuyoshi Yamaguchi

Abstract Screw-withdrawal tests were initially developed to be simple indicators of biological degradation. A quantitative test based on this premise was developed. Screw withdrawal resistance of a threaded probe inserted into a member similar to a wood screw can be examined, and a relationship between bending strength and corresponding probe resistance is known. It is confirmed by experiments that specific gravity, compression capacities, MOE and MOR of wood are correlated to screw withdrawal resistance of wood. Fundamental equations related between screw resistances and these physical values proposed by the researchers are summarized in this chapter. Reduction of specific densities is good indicators to evaluate the degradation of wood members for wood engineers, but compression capacity, MOE and MOR are better indicators for degradation of wood members for structural engineering.

Key words: screw withdrawal, density, MOE, MOR, compression

8.1 Background

Screw-withdrawal tests were initially developed to be indicators of biological degradation [1]. A quantitative test based on this premise was developed [2]. Screw withdrawal resistance of a threaded probe inserted into a member similar to a wood screw can be examined, and a relationship between bending strength and corresponding probe resistance is known [3] (see Figure 8.1).

B. Kasal and T. Tannert (eds.), In Situ Assessment of Structural Timber, 81–86.

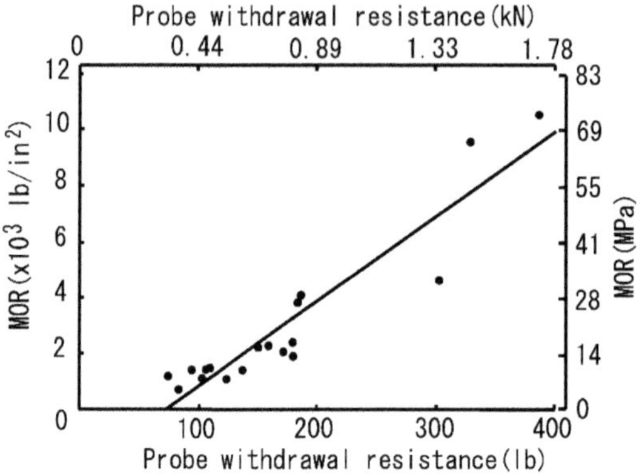

Fig. 8.1 Relationship between probe withdrawal resistance and residual bending strength

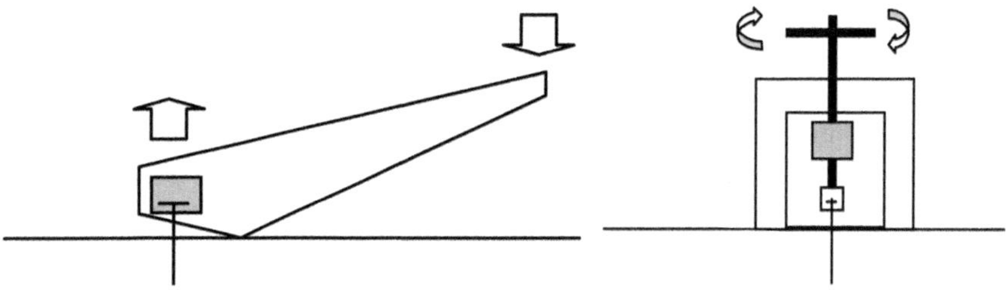

Fig. 8.2 Screw-extraction-force tool [1] (left) and clamp type tool [4] (right)

8.2 Equipment

Screw withdrawal resistances are measured by recording the required force to pull out a screw. There are mainly two types of tools to measure the withdrawal resistance. One is a screw-extraction-force tool, shown in Figure 8.2 left [1]. The load cell in the screw-extraction- tool measures the pull-out loads. The other tool is a clamp type tool shown in Figure 8.2 right [4]. Here the screw is secured in the clamp and pulled out by a fork. A load cell is connected between the fork and the shaft. The screw is pulled out by turning the handle, and the withdrawal resistance is measured by the load cell.

8.3 Application

8.3.1 Strength Estimate from Screw Withdrawal Test

Screw-withdrawal tests were applied for predicting bending strength of fire-retardant-treated plywood by Winandy et al. [1]. Multiple measurements per specimens were compared and single-measurements and two full-depth measurements were proposed to estimate the screw-withdrawal force per specimen. The effect of screw depth was evaluated. The half-depth measurements appeared to be less stable. For full-depth screw-withdrawal measurements, a general model was proposed:

$$\mathrm{MOR} = b_0 + b_1(\mathrm{SWF})^{1/2} + \mathrm{error} \tag{8.1}$$

where b_0 and b_1 are regression constants and SWF is screw withdrawal force.

8.3.2 Strength Estimate from Screw Withdrawals and Velocity of Stress Wave

Divos et al. [4] used screw withdrawal force and velocity of stress wave to propose the following empirical strength predictor equation for coniferous species:

$$\mathrm{MOR}_{\mathrm{est}} = 0.809 F_{\mathrm{screw}} \times v^2 + 26.8 \tag{8.2}$$

The applied units in the equations are: bending strength ($\mathrm{MOR}_{\mathrm{est}}$ [MPa]), F_{screw} [kN] and v [km/s]. The correlation coefficient between the bending strength and $\mathrm{MOR}_{\mathrm{est}}$ is 0.74. A similar MOR predictor formula applies for hardwoods:

$$\mathrm{MOR}_{\mathrm{est}} = 1.258 F_{\mathrm{screw}} \times v^2 + 36.9 \tag{8.3}$$

8.3.3 Screw Withdrawals and Angles of Entry

Gilfillan et al. [5] studied the relationship between screw withdrawals and angle of entry on fully threaded screw tests into clear timber as shown in Figure 8.3. The weakest pull-out forces occur when the screw is inserted parallel to the grain. Both the tangential and radial directions give similar results. A liner relationship is apparent between pull-out force and fully threaded screw depth for all three directions.

8.3.4 Screw Withdrawals and Compression Capacity

Gilfillan et al. [5] also studied the relationship between screw withdrawals and compression capacity (see Figure 8.4). The pull-out force values can be correlated with compressive strength.

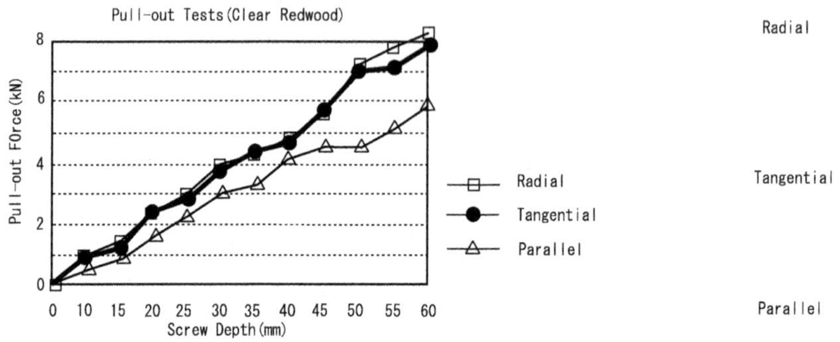

Fig. 8.3 Variation of pull-out force with angle of entry [5]

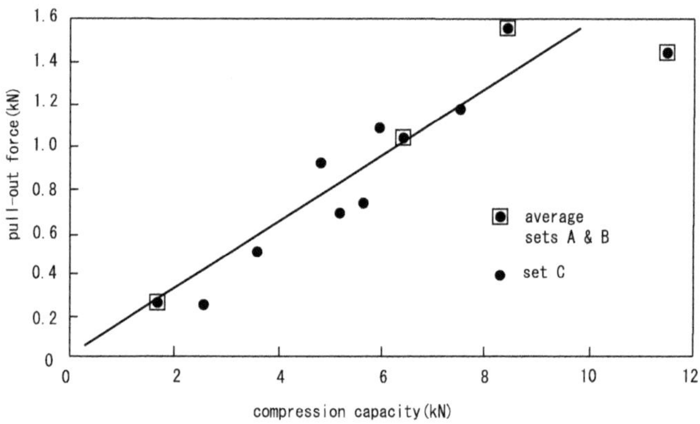

Fig. 8.4 Variation of pull-out force with compression capacity [5]

8.3.5 Densities from Screw Withdrawals

Screw withdrawal resistances of wood are correlated to wood specific gravities. The empirical equations obtaining screw withdrawal resistance from specific density are used in design specifications. Cai et al. [6] examined the relationship between screw withdrawals and density of wood. Specific gravity (G) was used as a response variable, and the maximum screw withdrawal load (P) was used as a predictor variable:

$$G = a + bP \tag{8.4}$$

where a and b are regression constants.

The regression results are shown in Figure 8.5. The screw withdrawal resistance method to determine density is relatively reliable and easy to apply.

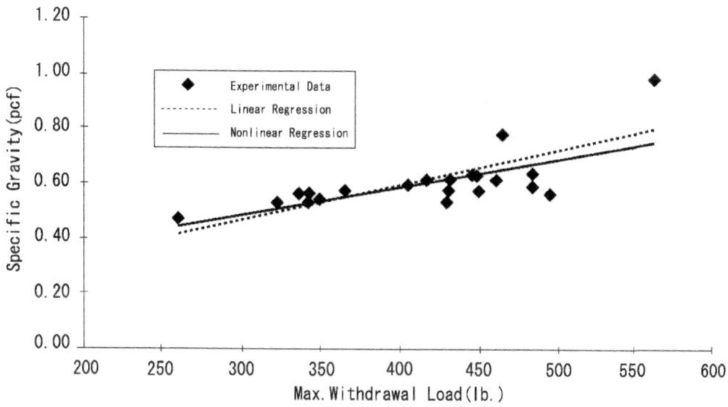

Fig. 8.5 Max. withdrawal load and specific gravity [6]

8.3.6 Stiffness Estimate from Screw Withdrawals and Velocity of Stress Wave

Cai et al. [6] also examined the relationship between screw withdrawals and dynamic modulus of elasticity (MOE). Dynamic MOE (E_d) were estimated using the following equation:

$$E_d = a + b\rho v^2 \tag{8.5}$$

where the density obtained from screw withdrawal testing, and the stress wave speed determined using stress wave timer, a and b are regression constants. The predicted MOE based on the average screw withdrawal loads is correlated to the static MOE [6].

8.3.7 Summary

It is confirmed by experiments that specific gravity, compression capacities, MOE and MOR of wood are correlated to screw withdrawal resistance of wood. Reduction of specific densities is good indicators to evaluate the degradation of wood members for wood engineers, but compression capacity, MOE and MOR are better indicators for degradation of wood members for structural engineering because structural engineers are required to evaluate damage of structures from the damage of members.

8.4 Limitations

Screw withdrawals are obtained from semi-destructive testing, when semi-destructive testing is not allowed in-situ, screw withdrawal resistances can not be obtained. Even if the screw withdrawal tests are allowed, the number of measuring points might be limited. Furthermore, the calculation of densities, compression capacity, MOE, MOR of members from screw withdrawals requires databases of regression constants for each parameters and each wood species. Only if wood screws or the probes are standardized, it will assist to build databases on screw withdrawal resistances. The effect of moisture contents, size effect, effect of sap wood and heart wood, etc. are not evaluated efficiently.

References

1. Winandy, J.E., Lebow, P.K., and Nelson, W. Predicting bending strength of fire-retardant-treated plywood from screw-withdrawal tests, Research Paper FPL-RP-568, Forest Products Laboratory, US Department of Agriculture, 1998.
2. Talbot, J.W. Unpublished research. Washington State University, Pullman, WA, 1982.
3. Ross, R.J. Nondestructive testing for assessing wood members in structures. General Technical Report FPL-GTR-70, Forest Products Laboratory, US Department of Agriculture, 1994.
4. Divos, F., Nemeth, L., and Bejo, L. Evaluation of the wooden structure of a Baroque palace in Papa, NDT Laboratory, University of Sopron, Hungary, woodndt.nyme.hu.
5. Gilfillan, J.R., Christie, D., and Gilbert, S.G. The residual strength of timber degraded by woodworm infestation. In: *Proceedings of Durability of Building Materials and Components 8*, Volume One, NRC Research Press, Ottawa, Canada, pp. 714–722, 1999.
6. Cai, Z., Hunt, M.O., Ross, R.J., and Soltis, L.A. Screw withdrawal – A means to evaluate densities of in-situ wood members. In: *Proceedings of the 13th International Symposium on Nondestructive Testing of Wood*, Forest Products Society, Madison, USA, pp. 277–281, 2002.

Chapter 9
Hardness Test

Mariapaola Riggio and Maurizio Piazza

Abstract The hardness of materials is commonly characterized using indentation techniques. The hardness values depend on the tool employed and the measured parameter, for wood, the anisotropy, heterogeneity and hygroscopicity also influence the measurements. In this chapter, different methods used for hardness testing of wood are presented. Possible correlations between the different hardness figures and other mechanical properties are indicated, in order to highlight the suitability of a specific technique for the assessment of structural timber on site. Limitations of each method are also discussed. A first group of testing methods is based on the static macroindentation, among them, the Janka, Brinell and Monnin hardness tests are standardized procedures, but they are not specifically designated to test structural timber for the estimation of mechanical properties correlated with hardness. The Piazza and Turrini hardness test, on the contrary, is conceived as a NDT method for the mechanical characterization of timber elements. Hardness tests of wood based on dynamic indentation, are briefly described: they are commonly used to map decay in wood, rather than to estimate global mechanical properties of structural elements.

Key words: Brinell, density, hardness, indentation, Janka, Monnin, specific gravity

9.1 Background

Hardness refers to properties of solid materials that give them resistance when a force is applied. In materials science, there are three principal definitions of hardness: (i) indentation hardness: resistance to plastic deformation due to a constant load; (ii) rebound hardness: height of the bounce of an object dropped on the material, related to elasticity; and (iii) scratch hardness: resistance to fracture or plastic deformation due to friction from a sharp object.

The hardness of materials is commonly characterized using indentation techniques because of the ease and speed of conducting tests. Properties that are meas-

B. Kasal and T. Tannert (eds.), In Situ Assessment of Structural Timber, 87–97.

ured by indentation describe the deformation of the volume of material beneath the indenter. The deformation that occurs during indentation can involve elasticity, viscoelasticity, plasticity, creep and fracture and are described by the elastic modulus, relaxation modulus, hardness, creep rate, and fracture toughness respectively. Characterizing these properties provides a method for describing a material's response to an applied loading condition, which helps predict material performance.

Anisotropy, heterogeneity and hygroscopicity of wood influence the measurements. The hardness values also depend on the tool employed [1]. The use of different indenter geometries, such as ball, wedge, conical, pyramidal or cylindrical tools, involves different phenomena such as friction, cleavage and shearing. Friction, for example, as well as average strain beneath the indenter, is greater with a sharp tool than with a blunt tool. Also the dimension of the indenter influences the test results; in case of macro-indentation, the size of the tool has to be comparable with the scale of heterogeneities in the wood structure; therefore, average values of properties at the macro-scale are generally measured with tools whose size encompasses several growth layers.

Indentation tests, in general, differ not only in shape and size of the indenter but also according to the measured parameter. By using a fixed load and measuring the resulting diameter or width of the impression at the surface; by using the contact area in computing the mean unit load on the indenter (Brinell hardness) [2], or by using the projected area of the impression on the surface (Meyer hardness); by using a fixed load and measuring the resulting depth of impression; and by using a variable load to produce a given depth of impression.

Hardness measurements further depend on the surface orientation. Wood hardness is higher on the cross-section face (end-hardness) than on the longitudinal faces (side-hardness). Mörath [3] presented hardness values for radial, tangential and cross sections. He found that the difference in hardness between cross and side sections decreased with increasing density. Radial hardness is higher than tangential (about 5–10%) in hardwoods with high amount of wood rays, as e.g. in beech or oak [4]. Indentation hardness tests can be performed on a macroscopic or microscopic scale, as well as applying the loading force statically or dynamically.

9.2 Equipment and Application

Different methods are used for hardness testing of wood. A first group of testing methods is based on the static macroindentation, among them, the Janka, Brinell and Monnin hardness tests are standardized procedures, but they are not specifically designated to test structural timber for the estimation of mechanical properties correlated with hardness. The Piazza and Turrini hardness test, on the contrary, is a method for the characterization of timber.

Hardness tests of wood can be also based on the dynamic indentation. Techniques based on the dynamic penetration of a thin indenter, are commonly used to map decay in wood, rather than to estimate global mechanical properties of structural

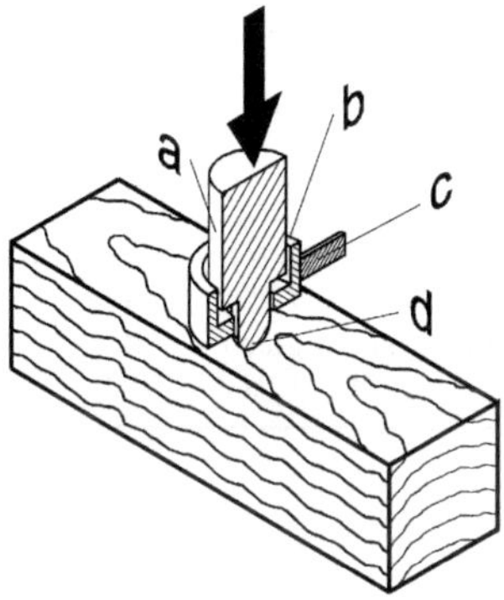

Fig. 9.1 Historical equipment for ASTM D143 test of Janka hardness: (a) shaft of test jig; (b) flexible collar; (c) level for collar; (d) ball indenter

elements. Microndentation tests can be used to probe the mechanical properties at the cell and sub-cell wall level [5]. Because of the difficulties in predicting properties of wood at larger scales, microhardness is especially adopted in biological-based research, rather than in non-destructive testing of structural timber.

9.2.1 Janka Hardness Test

Janka [6, 7] implemented a modified Brinell hardness test (1900) for wood, based on the force required by static loading to embed a steel hemisphere with a diameter of 0.444 in (projected area of 100 mm^2) completely into the wood. The Janka hardness test was adopted by the American Society for Testing and Materials (ASTM). Although originally expressed by Janka as a load divided by the projected area of contact, the ASTM D143 [8] hardness value has always been specified as the load (H) at a penetration of 0.222 in. The ASTM D143 test is conceived for laboratory testing. The standard specimen is a solid piece of wood with a cross section of 2 by 2 in and length of 6 in. A distinction is made between hardness determined on the end and on the side of the piece. No distinction is made between hardness on the radial and tangential surfaces. The test equipment originally had a collar to which the ball was attached; the ball was penetrated into the specimen until the collar was tightened against the specimen (Figure 9.1). In 1948, an electronic circuit indicator was added for determining the penetration.

Janka hardness was initially used as a minimally destructive test for estimating the strength and stiffness properties of wood [8]. Currently side hardness, determined with the Janka test, is used to assess the suitability of wood species for use as flooring. Regarding the capability of Janka hardness to be an indicator of wood strength properties, correlations were found by some authors; however, these are only valid for clear wood at specific moisture content. From the results of hardness tests on 280 wood species, Janka found an empirical relationship between hardness

Table 9.1 Relationship between Janka's hardness and specific gravity [10]

Species	MC	$H_j = A\rho^n$	
		A	n
Hardwood	green	3720	2.31
	12%	3400	2.09
Softwood	green	1400	1.41
	12%	1930	1.50

and crushing strength:

$$H_j = 2\sigma_{cb} - 500\ [\text{kp/cm}^2] \tag{9.1}$$

In an unpublished report, Pettigrew and Newlin [8] indicated, for green softwoods, the following correlation between Janka hardness and modulus of rupture (MOR):

$$\text{MOR} = 72H_j^{(3/4)}\ [\text{lb/in}^2] \tag{9.2}$$

The relationship between Janka hardness and ultimate compressive stress (UCS) is:

$$\text{UCS} = 35H_j^{(3/4)}\ [\text{lb/in}^2] \tag{9.3}$$

While MOR of hardwoods generally increases with increasing hardness, this relationship was considered "very indefinite" and "an attempt to determine strength from hardness with any degree of accuracy would be useless" [9]. The relationships are only valid for green clear wood; therefore, they are not directly applicable for grading structural timber.

Hardness can be also used as an indicator of density. Janka found that hardness is approximately proportional to the density of the wood. Based on numerous measurements, Newlin and Wilson [1] determined that the relationship between hardness and specific gravity (ρ) may be expressed as a power formula and derived the following equation:

$$H_j = A\rho^n\ [\text{kp/cm}^2] \tag{9.4}$$

Newlin and Wilson gave separate coefficients, A and n, for green and dry wood, respectively, but did not separate hardwoods from softwoods. Relationships for Janka hardness and specific gravity for both green and dry hardwoods and softwoods are provided in the *Wood Handbook* [10] (Table 9.1).

9.2.1.1 Brinell Hardness Test

An application of the Brinell hardness test to wood was proposed by Mörath in 1932 [2]. Brinell hardness is measured from the diameter d of the impression of a steel ball, 10 mm in diameter, after the application of a normal load of 50 kp (for extremely hard species 100 kp and for very soft ones 10 kp). The maximum load P should be reached within 15 sec, kept constant over a period of 30 sec, and than

Table 9.2 Values for the constants a and b used in Equation (9.6) [1]

	End-hardness		Side-hardness	
	a	b	a	b
European woods	1180	1.62	670	2.14
American woods	1200	1.53	680	2.0

Table 9.3 Values for the constants α and β used in Equation (9.7) [1]

End-hardness		Side-hardness	
α	β	α	β
–213.3	1263.3	–145.4	664.2

reduced to zero within another 15 sec. If P is the maximum load, D the diameter of the steel ball, and d the diameter of the impression, the value of the Brinell hardness H_B can be calculated as follows:

$$H_B = \frac{2P}{\pi \cdot D \cdot (D - \sqrt{D^2 - d^2})} \ [\text{kp/cm}^2] \tag{9.5}$$

The calculation of hardness, based on load and diameter of impression is unreliable (Pallay in [1]). A steel-ball with 10 mm diameter is too small with respect to the heterogeneous structure of wood. The Brinell hardness test is the standard test for the determination of the resistance to indentation of flooring [11]. The relationship between Brinell hardness and specific gravity (ρ) may be expressed according to Trendelenburg (in [1]) as

$$H_B = a \cdot \rho_0^b \ [\text{kp/cm}^2] \tag{9.6}$$

where ρ_0 is the oven-dry, and a and b are constants given in Table 9.2.

A linear relationship between Brinell hardness and oven-dry density is applicable [12]:

$$H_B = \beta \cdot \rho_0 + \alpha \ [\text{kp/cm}^2] \tag{9.7}$$

where α and β are constants given in Table 9.3.

9.2.1.2 Monnin Hardness Test

In France, the test according to Chalais–Meudon, also known as Monnin hardness test (Figure 9.2), is standardized [13]. In the Monnin test, a 30-mm diameter steel cylinder is impressed with a maximum load of 200 kp (2 kN) on a radial section over a period of 5 sec. Since it is difficult to measure the depth of penetration t, it is deduced from the width l of the impression

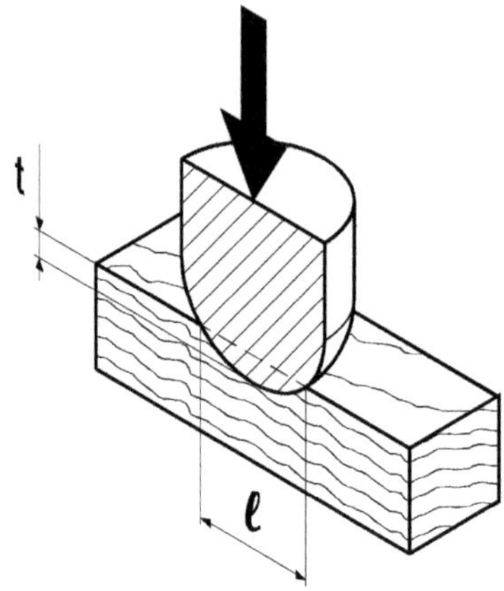

Fig. 9.2 Monnin hardness test: indentation measurement

$$t = 15 - \frac{1}{2}\sqrt{900 - l^2} \tag{9.8}$$

The Monnin hardness is then defined as the reciprocal of the depth of penetration:

$$H_M = \frac{1}{t} \tag{9.9}$$

Because it is not easy to measure accurately the width of the impression [14], the Monnin hardness is subject to greater experimental error than in the Janka test. However, it has merits for high density timbers where the Janka tool induces splitting [15].

9.2.1.3 Piazza and Turrini Hardness

The hardness test proposed by Piazza and Turrini [16] is a modified Janka test that measures the force R required to embed a 10 mm steel bit to 5 mm. The experimental test equipment is shown in Figure 9.3.

In order to estimate the global behaviour of a structural element the value of R must be obtained by averaging the test results made on the longitudinal faces of the element. Each test consists in five measurements taken in a limited portion of the element. The result is obtained by averaging three median values among the five measures. The test area must be clear, without visible defects. The method, however, is specifically conceived for assessing structural timber on site, and a correction factor δ is introduced (see Equation (9.10)), that correlates R and Young's modulus E_0 in longitudinal direction, in order to take into account the presence of defects:

$$E_0 = \delta \cdot A \cdot R^{0.5} \text{ [MPa]} \tag{9.10}$$

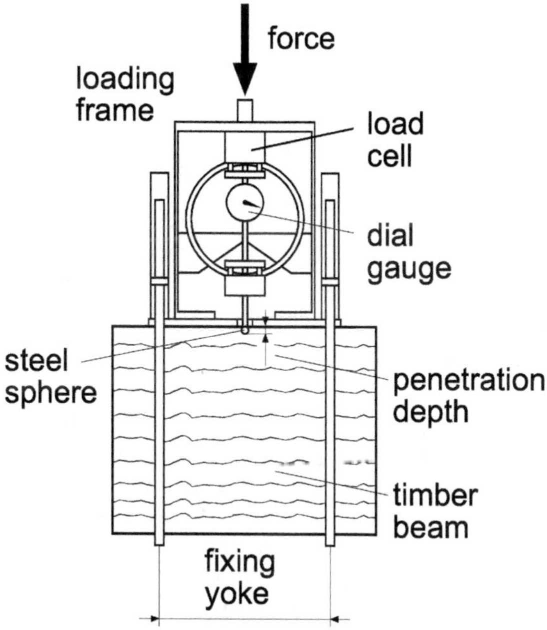

Fig. 9.3 Hardness test device [16]

where A is a coefficient that depends on the species of the tested sample, the value of the applied force and the moisture content.

For Silver fir (*Abies Alba*) and Larch (*Larix decidua*) at a moisture content in the range of 12–14%, $A = 350$ and for Chestnut (*Castanea sativa*) a value of $A = 263$ was proposed [17, 18]. The experimental force ranges from 700 to 3000 N and is calibrated according to the loads in service (stress values equal to 60–70% of the failure load). Hence, the hardness is calculated from the linear part of the force-penetration plot. For structural timber, the reduction factor δ is in the range of 0.5–0.8, depending on the defectiveness of the element. Three separate values of δ are proposed according to the presence and size of defects measured on the visible faces of the timber element (Table 9.4). Criteria for the measurements of the defects are deduced from the UNI 11119:2004 [19], for the visual grading of structural timber on site.

In order to adjust the value of E_0 for different values of moisture content, the result must be divided by a correction factor C [17, 18]:

$$C = (1 - 0.0079\Delta_{U15-u[\%]}) \tag{9.11}$$

where $\Delta_{U15-u[\%]}$ is the difference (%), between the MC at 15% and the test MC.

Table 9.4 Values of δ according to the size of defects

Characteristics	δ		
	0.5	0.68	0.8
Single knots	$\leq 1/5$ ≤ 50 mm	$\leq 1/3$ ≤ 70 mm	$\leq 1/2$
Group of knots	$\leq 2/5$	$\leq 2/3$	$\leq 2/5$
Slope of grain	$\leq 1/14$	$\leq 1/8$	$\leq 1/5$
Checks	/	/	limited

9.2.1.4 Wedge Hardness

A wedge indenter is a triangular prism. Therefore, wedge hardness calculated from the projected area of the indentation, is a function of tool angle as this determines the degree of confinement of the wood tissue. The use of wedges that are longer than the specimen width has been advocated [20]. Measurements with the wedge, as well as with the cylinder, must be taken with the tool edge lying parallel to the radial direction as for the Monnin test, in order to intercept both earlywood and latewood layers. Doyle and Walker [20] give the following relation between wedge hardness and density:

$$H_w = y_0 + y_1 \rho + y_2 \rho^2 \tag{9.12}$$

where ρ is the wood density and y_n are constants, whose values are, for the wedge's angle of 136° apical angle:

$$y_0 = -1.545; \quad y_1 = 0.0122; \quad y_2 = 0.00004 \tag{9.13}$$

9.2.1.5 Cone Hardness

Hardness measured with conical or pyramidal indenters, rather than with spherical and cylindrical tools, is independent of the depth of indentation, being a function of the confining pressure and directly relatable to the tool angle. For measurements based on the projected contact area tools with a rounded tip, as in the Rockwell test [8] are recommended [20]. Kumichel and Holz (in [1]), using a slightly truncated 530 cone, obtained a more extended scale of hardness with wood density and less variance within sample groups as compared to the Brinell ball.

9.2.1.6 Dynamic Indentation

Baumann (1922), Schwarz and Bues (1929), Gaber (1935) and Pevzoff (1935) (in [1]) developed dynamic hardness tests by dropping a steel-ball on the wood surface. They found a strong correlation between dynamic end-hardness and crushing strength along the grain. A typical dynamic indentation test on timber elements

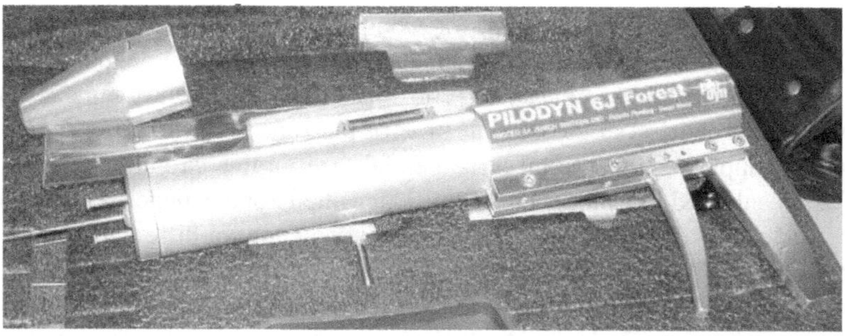

Fig. 9.4 Pilodyn device

uses a slender steel pin of a given diameter, which is driven at a constant energy into the wood surface by releasing a compressed spring. A commercial device based on this technique is the Pilodyn: it shoots a steel rod 2.5 in diameter against the tested surface with impact energy depending on the stiffness of the spring of individual devices (Figure 9.4). The typical range of measurements covers depth between 0 and 40 mm and the results is displayed on the scale of the device.

Görlacher [21] obtained good correlation coefficients between density and depth of penetration of the Pilodyn 6J, taking into account that the number of measurements for each specimen must be large. A proposed empirical relation between the depth of penetration and density was affected by moisture content. The correlation coefficient varied from 0.74 to 0.92, and depended on number of measurements and species, therefore, species-based calibrations are required. Average Pilodyn data obtained from the whole element show weak correlation with the mean density of timber elements [22]. On the contrary, single Pilodyn data permit mapping the condition of the surface decay and highlight the regions of the element that need further investigation with more sophisticated NDT methods.

9.2.1.7 Limitations

Hardness testing in wood is more difficult and less reliable than testing isotropic and homogenous materials, such as metals. Wood hardness involves compression strength, shear strength and fracture toughness and results are also influenced by friction and cleavage. The correlated properties are dependent on the parameters of the hardness test (shape of indentation tool, speed of loading and depth of penetration) and the way in which wood failure is induced during testing. Therefore there is no simple relationship between the results of different hardness tests. Wood shows qualitatively different behaviors under the various measurement methods.

Common problems occur with methods based on the measure of the impression, especially of shallow indentations where the imprint is indistinct. Sawada et al. [23] observed that because elastic modulus to yield stress ratio (E/Y) perpendicular to the grain is low (E/Y of wood lies in the range 20–60, against 100–1000 of metals) elastic recovery is great for wood, especially if green, and makes the measurement of the size of the permanent indentation of questionable value. Also the "sinking in" phenomenon causes problems in determining the actual size of the indentation. It consists in the densification of the zone beneath the indenter, when the load is applied on the radial or tangential face; in this case the densified zone acts as an enlarged indenter. The size of a densified zone is dependent on the tool shape and causes different resistance against indentation for different tools [12, 15]. Other sources of uncertainties in the measure are introduced by the machine stiffness, the surface roughness of the specimen and the eccentric initial contact between specimen and tool, typical when tools with a single axis of symmetry are used.

Because wood is a hygroscopic material, relations between hardness and timber moisture content must be known. Below the fiber saturation point, hardness is in inverse proportion to moisture content. For softwoods, hardness values of air dry samples ($u \approx 12\%$) are 1.7–1.8 higher than for green samples ($u > 35\%$). For moisture content values within the range of approximately 6 to 20%, change in hardness (H) with change in moisture content (M) may be estimated from the following formula [10]:

$$H = H_{12}(H_{12}/H_{\text{green}})^{[(12-M)/(M_p-12)]} \tag{9.14}$$

where H_{12} is hardness at 12% moisture content, H_{green} is hardness of green lumber, M moisture content (%), and M_p is species dependent intersection moisture content (range 21–28%).

All the experimental relations between hardness and other mechanical properties of wood can be used just within the experimental limits of each method (species, moisture content, loading rate). Standardized methods, that are conceived for tests of clear wood specimens, of a specific size and shape, are of poor benefit to the assessment of structural timber on site, if no considerations on the presence of defects are introduced.

References

1. Kollmann, F.F.P. and Côté, W.A. *Principles of Wood Science and Technology*, Vol. I. Springer Verlag, 1984.
2. Brinell, J.A. Ein Verfahren zur Härtebestimmung nebst einigen Anwendungen desselben. *Gießlers Baumaterilienkunde*, Vol. 5, 1900.
3. Mörath, E. Studien über die hygroskopischen Eigenschaften und die Härte der Hölzer. Habilitationsschrift zur Erlangung der Venia Legendi für Technologie des Holzes der Abteilung an der Technischen Hochschule Darmstadt, 1932.
4. Görlacher, R. Untersuchung von altem Konstruktionsholz: Bestimmung des Elastizitäts-Modulus. *Bauen mit Holz*, Vol. 8/91, pp. 582–586, 1991.
5. Tsui, T.Y. and Oliver, W.C. Longitudinal hardness and Young's modulus of spruce tracheids secondary walls using nanoindentation techniques. *Wood Science and Technology*, Vol. 31(2), pp. 131–141, 1997.
6. Janka, G. Die Härte des Holzes. *Zentralblatt für das gesamte Forstwesen, Wien*, Vol. 32(5), pp. 193–202, 1906.
7. Janka, G. Die Härte des Holzes. *Zentralblatt für das gesamte Forstwesen, Wien*, Vol. 32(6), pp. 241–260, 1906.

8. ASTM E 18-79, Standard test methods for Rockwell hardness and Rockwell superficial hardness of metallic materials. ASTM International, 1981.
9. Green, D.W., Begel, M., and Nelson, W. Janka hardness using nonstandard specimens. Res. Note FPL-RN-0303. US Department of Agriculture, Forest Service, Forest Products Laboratory, Madison, WI, 13 pp., 2006.
10. Forest Products Laboratory. *Wood Handbook – Wood as an Engineering Material.* US Department of Agriculture, Forest Service, Forest Products Laboratory, Madison WI, 1999.
11. EN 1534:2000. Wood and parquet flooring – Determination of resistance to indentation (Brinell) – Test Method. European Committee of Standardization (CEN), 2000.
12. Ylinen, A. Über den Einfluß der Rohwichte und des Spätholzanteils auf die Brinellhärte des Holzes. *Holz Roh- Werkstoff*, Vol. 6(4), pp. 125-127, 1943.
13. AFNOR NF B 51-125. Bois-essai de dureté, “Monnin”. L’Association Française de Normalisation (AFNOR), 1972.
14. Sunley, J.G. A comparison of the Janka and Monnin methods of testing the hardness of timber and wood products. *J. Inst. Wood Sci.*, Vol. 14, pp. 40–46, 1965.
15. Doyle, J. and Walker, J.C.F. Indentation hardness of wood. *Wood and Fibre Science*, Vol. 17(3), pp. 369–376, 1984.
16. Piazza, M. and Turrini, G. Il recupero dei solai in legno. Esperienze e realizzazioni. *Recuperare*, Vol. 7, 1983.
17. Piazza, M. and Riggio, M. Visual strength grading of traditional timber beams. Limits of the method: The old roof beams of the “Ai caduti dell’Adamello” refuge, In: *Proceedings of Structural Faults + Repair-2006 Conference*, Edinburgh, UK, 13–15 June 2006.
18. Piazza, M. and Riggio, M. Visual strength grading and NDT of timber in traditional structure. *Journal of Building Appraisal*, Vol. 3, pp. 267–296, 2008.
19. UNI 11119. Cultural heritage – Wooden artifacts – Load-bearing structures – On site inspections for the diagnosis of timber members. Ente Nazionale Italiano di Unificazione, UNI, Milano, Italy, 2004.
20. Doyle, J. and Walker, J.C.F. Indentation of wood by wedges. *Wood Science and Technology*, Vol. 19(1), pp. 47–55, 1985.
21. Görlacher, V.R. Zerstörungsfreie Prüfung von Holz: Ein in Situ Verfahren zur Bestimmung der Rohdichte. *Holz als Roh- und Werkstoff*, Vol. 45(7), pp. 273–278, 1987.
22. Piazza, M. and Riggio, M. NDT methods for the assessment of structural timber: report on the researches carried out at the University of Trento (Italy). In: *On Site Assessment of Concrete, Masonry and Timber. SACoMaTIS 2*, pp. 1039–1048, 2008.
23. Sawada, M., Tsujl, K., and Kondo, K. Relationship of hardness to compressive strength of wood. Report No. I. *Ringyo Shikensho Kenkyn Hokoko*, Vol. 78, pp. 149–174, 1955.

Chapter 10
Moisture Measurement

Bohumil Kasal and Gretchen Lear

Abstract Moisture content (MC) of wood affects almost all physical and mechanical properties of wood and moisture content must always be measured. Indirect methods using relationship between some physical quantities such as electrical resistance and MC are frequently used. Such methods usually require calibrations for various species, temperature corrections and are direction dependent. Direct (weight method) method will give accurate moisture content measurement but both, indirect and direct methods are local and number of measurements is required to obtain MC value representing the average MS of investigated structural member.

Key words: moisture contents, electrical resistance, calibration, oven drying, pin meter, pin-less meter, electrode, mechanical properties

10.1 Background

Elevated moisture content of members can affect the results of nondestructive and semi-destructive testing, particularly stress wave behavior. Moisture content (MC) also has a surprisingly strong influence on mechanical properties and should be considered when inspecting in situ members. Moisture infiltration in structural systems is a common source of damage to structural materials and moisture measurements can help to identify water infiltration paths as well as locate areas that may need further investigation to assess damage and material condition associated with elevated moisture exposure.

These factors should be considered during in situ assessments and moisture measurements can be used to aid in nondestructive and semi-destructive investigations.

B. Kasal and T. Tannert (eds.), In Situ Assessment of Structural Timber, 99–104.

10.2 Equipment

Oven drying of wood specimens is the traditional method of measuring moisture contents; however this method is not suitable for in situ investigations and therefore hand held moisture meters are commonly used. Substantial data for wood exists to calibrate moisture meters, allowing moisture content to be quantified from meter readings based on wood species and temperature [1]. Quantified moisture readings represent the water weight in the member as a percentage of the oven dry weight of the member.

Moisture meters can be described and grouped by the principles on which they operate; resistance, capacitance, voltage, microwave, or thermal methods. Resistance methods utilize the electrical resistance of the material, capacitance methods measure radio-frequency power loss, and voltage methods measure moisture in terms of a DC voltage across a known resistor. Microwave methods also utilize radio-frequency power loss but at higher frequencies, and thermal methods are based on temperature change associated with moisture content changes [1]. In situ member assessments are generally intended to evaluate current structural conditions and are not intended as long term monitoring exhibitions; therefore hand held moisture meters are most often employed and discussion will be limited to these devices.

Hand held moisture meters utilize the electric properties of materials to measure MC and are most commonly of the resistance and capacitance type. Resistance-based moisture meters are employed in probe-type or pin meters while capacitance methods are applied with surface or pin-less meters.

10.2.1 Pin Meters

Pin meters, those which operate on resistance principles, are commonly employed for in situ timber investigations. Pins are driven into the member and the resistance to electrical current is measured between them. Electrical resistance of material will vary depending on the moisture content. As moisture content increases, the resistance of the material reduces and the conductance increases. Water is a good conductor and wood is a poor conductor of electrical current, therefore the effect of the moisture content on the resistance properties makes pin meters well suited for timber investigation.

Pin length will vary depending on manufacturer and can be as long as two inches, allowing for the investigation of thick members. Pins should be inserted into the member so that they and the current flow are parallel to the grain. If oriented perpendicular to the grain the current will encounter more resistance as it crosses grain boundaries, which will be reflected in the moisture readings [2].

Meters with insulated pins measure moisture content at the depth of the pin penetration since the electric current only flows and the resistance is only measured in between the un-insulated tips of the pins. This allows for collecting moisture content readings at multiple depths to evaluate moisture gradients. In contrast, un-insulated pin meters do not allow for moisture content readings at specified depths. Electric current will follow the path of least resistance, i.e. the wettest layer penetrated by the pins; therefore the reading will represent the wettest portion of wood [2, 3].

The accuracy of pin meters is limited at low moisture contents. Below 6 or 7% moisture content, the effect of moisture on the member's resistance properties is comparable to that of wood alone and accurate measurements of moisture are not attainable. The fiber saturation point is the upper limit range on readings from pin-meters; although conductivity of wood increases as the moisture increases, past the fiber saturation point the increase is much smaller and erratic, eliminating the ability to take reliable readings. Increased temperatures will increase conductance and published correction factors should be applied to readings on specimens with temperature over 90°F (32°C) or below 70°F (21°C) [3].

Moisture readings of wood materials can be affected by species, grain distribution, temperature, chemical treatments, as well as the skill of the operator. These sources produce variability in readings. Standard deviation for resistance meters is from 0.5 to 1.5 percentage points, meaning that resistance meter readings at their best will be off by one percentage point about 5% of the time [3].

10.2.2 Pin-less Meters

Pin-less meters, operating on dielectric principles, operate at the surface of the timber. A frequency signal is sent into the timber by transmitting electrodes and received by surface contact electrodes. Attenuation of the frequency signal indicates the degree of moisture contents [1]. Although less precise than pin meters, pin-less meters can estimate moisture content levels down to 0% [3]. At elevated moisture contents, the pin-less meter can give some estimates on moisture content; however the fiber saturation point is still the upper reliable limit.

Since pin-less meters do not penetrate the member they are more susceptible to influence by the material near the surface. Material more than approximately 0.32 mm (0.125 in) below the surface is not properly weighted in the moisture reading [3]. Temperature affects dielectric properties, in a more complex manner than it does on conductance based readings. Adjustments should be made accordingly with reference to published charts, examples of which can be found in [3].

Pin-less meters are also influenced by factors other than moisture content, species, grain distribution, temperature, etc., which cause some variance in reading. Standard deviation for dielectric meters is from 1–3%; in other words, the readings would be expected to have an error of 1% approximately 15% of the time [3].

Table 10.1 Percent loss of mechanical properties due to early decay [4]

Strength Property	Average strength loss due to decay in percent
Static bending	70
Impact bending	80
Modulus of elasticity	70
Compression parallel to grain	45
Tension parallel to grain	60
Compression perpendicular to grain	60
Shear	20

10.3 Application

Moisture content readings can identify areas in members, which have elevated moisture levels that create the environmental conditions necessary to support biotic deterioration. Recognizing areas prone to biotic attack allows for further investigation to identify and quantify any existing damage. Biotic deterioration refers to the damage caused by the attack of living agents such as bacteria, fungi, or insects. Each of these has multiple species that can often cause severe damage or total destruction of a timber member. Each requires life sustaining conditions; oxygen, tolerable temperatures, a food source (in this case wood) and moisture.

Elevated MC that range beyond the fiber saturation point and allow for free or unbounded water foster the growth of decay fungi that destroy timber fibers and weaken the member. Typically, moisture contents below 20% will not harbor fungal growth. Elevated moisture within a member will also increase the chance of insect attack and is, according to Ross et al. [9], the most significant contributing cause of insect attack. Infestation can also attract insect hunting birds, such as woodpeckers, whose damage to the timber can be greater than the insects'.

Decay from these living organisms can substantially reduce material properties. Mechanical properties can be reduced by 10% before visual indicators are present. With a weight loss of only 5–10%, the loss in mechanical properties can be as large as 80%. Listed in Table 10.1 is the probable strength loss associated with early decay of softwood [4], (only a demonstration of the possible effect of decay on mechanical properties).

In order to determine the remaining strength of members, additional research would be required; however it is clear that decay can greatly inhibit a member's load resistance ability.

Moisture content strongly influences mechanical properties of timber members as well as causes dimensional changes. In general, elasticity and strength parameters decrease with increased moisture content up to the fiber saturation point, as illustrated by Figure 10.1. The effect of moisture content on mechanical properties is also linked to the quality and size of the timber as shown in [5, 6]. Adjustments are made to mechanical properties to account for moisture content differences, and it is commonly assumed that the strength variation below the fiber saturation point can be described by a negative exponential function [8]. Extensive discussion on this

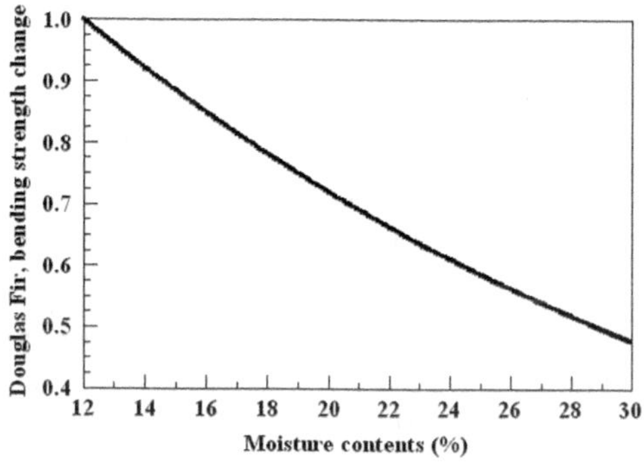

Fig. 10.1 The effect of moisture content on mechanical properties [8]

topic and the effects of moisture on the mechanical properties of timber beyond this scope can be found in [7].

10.4 Limitations

Meters can be affected by parameters other than the moisture content including timber species, grain distribution, temperature, chemical or preservative treatments, as well as the skill of the operator [3]. Limitations based on meter type also exist. Pin meters have the drawback of leaving holes in the specimen as well as the inability to measure low moisture contents. They are also more sensitive to temperature factors than the pin-less meters and require corrective changes. Pin-less meters do not alter the material surface but good contact to the surface, which can be difficult with rough or warped members, is required for accurate readings. Pin-less meters are also more sensitive to density differences and generally require correction or settings specific to the member. Additionally, the lack of penetration into the member makes pin-less meter's readings biased to surface material conditions.

References

1. Said, M.N. Moisture measurement guide for building envelope applications. Research Report, Institute for Research in Construction, National Research Council Canada, Vol. 190, pp. 1–34, 2004.
2. Hoadley, R.B. *Understanding Wood: A Craftsman's Guide to Wood Technology*. Taunton Press, Newtown, 2000.
3. James, W.L. Fundamentals of hand held moisture meters: An outline. In: *Proceedings of ASTM Hand-held Moisture Meter Workshop*, Madison, WI, 5 May 1993. Forest Products Society, Madison, WI, pp. 13–16, 1994.
4. Stalnaker, J.J. and Harris, E.C. *Structural Design in Wood*, 2nd ed. Kluwer Academic Publishers, Boston, 2002.

5. McLain, T.E., DeBonis, A.L., Green, D.W., Wilson, F.J., and Link, C.L. Pine dimension lumber. Research Paper FPL 447. US Department of Agriculture, Forest Service, Forest Products Laboratory, Madison, WI, 40 pp., 1984.
6. Green, D.W., Evans, J.W., and Pellerin, R. Moisture content and the flexural properties of lumber: Species differences. In: *Proceedings of International Timber Engineering Conference*, Vol. 2. TRADA, London, UK, pp 2.181–2.188, 1991.
7. Bodig, J. and Jayne, B. *Mechanics of Wood and Wood Composites*. Van Nostrand Reinhold, New York, 1982.
8. Kasal, B. Mechanical properties of wood. In: *Encyclopedia of Forest Science*, J. Burley, J. Evans, and J. Younquist (Eds.), Elsevier, Oxford, England, pp. 1815–1828, 2004.
9. Ross, R.J., Brashaw, B.K., Wang, X., White, R.H., and Pellerin, R.F. *Wood and Timber: Condition Assessment Manual*. Forest Products Society, Madison, WI, 73 pp., 2004.

Chapter 11
Species Identification

Nicola Macchioni

Abstract The chapter deals with methodology to be followed for a reliable identification of the wood species present in timber structures. The determination of the species is basic information for any analysis on wooden artefacts. Specifically for in situ timber structure diagnosis, the determination of the species is crucial for the correct determination of the structural performances of each element. The procedure for the identification of the species passes through three steps: sampling, macroscopic evaluation, microscopic evaluation. A correct sampling, in terms of sample number, representativity and dimensions allows a correct identification procedure; the macroscopic evaluation, with or without the help of a magnifying glass, examines the characteristic aesthetical features (colour, vein, texture) of the species, but rarely allows a reliable identification of the species. Then the final step is the microscopic evaluation that examines the anatomical characteristics of the sample by means of optical or electronic microscopes. Both macroscopic and microscopic identification require significant training experience. In some cases it may be impossible to identify the species and it might be necessary to terminate the identification at a level higher than the species within the botanic taxonomy.

Key words: hardwood, macroscopic identification, microscopic identification, species identification, microscope, sampling, softwood

11.1 Equipment

The identification of species or species group is critical for evaluation of physical and mechanical properties of in-situ members and is required in any in-situ investigation. Species identification uses comparative techniques. The first stage involves macroscopic evaluation. When typical macroscopic (aesthetic) characteristics are partially or totally absent or are inconclusive, the microscopic technique must be used.

B. Kasal and T. Tannert (eds.), In Situ Assessment of Structural Timber, 105–107.

The sampling stage is critical for a reliable identification, but it requires physical removal of a part of the object, which may be viewed as too destructive from a conservation viewpoint. The sample(s) must satisfy requirements of pertinent standards such as [1]:

- be sufficient in number: the sampling must be done from each member that, after macroscopic observation, appears to belong to different species;
- be three-dimensional and include the three anatomical directions of wood;
- be large enough to obtain the three sections (radial, tangential and transverse surfaces) required for reliable identification: the dimension along the three principal directions can be several millimeters.

11.1.1 Macroscopic Evaluation

Macroscopic evaluation, generally performed without mechanical assistance or with the help of a magnifying glass, must be conducted on a surface that is free of contamination. This is typically done by removing a surface layer of the sample via a sharp blade.

This type of identification is done through the observation of wood macroscopic features along the three principal wood directions (transversal, longitudinal radial and longitudinal tangential); color (sapwood/heartwood); size of the growth rings on transverse and/or longitudinal planes (figure, vessels, texture), and the shape and size of parenchymatic rays on transverse and longitudinal surfaces [3].

As mentioned above, this is a comparative process where the unknown specimen is compared with standards defined in a reliable xylarium or a scientific atlas of wood species. This requires a significant level of training and experience in wood anatomy.

11.1.2 Microscopic Evaluations

Frequently in old timber structures, the macroscopic features are hidden by surface contamination, oxidation, and/or surface treatment (decorations, e.g.) that prevents macroscopic species identification. Proper sampling for microscopic evaluation and identification becomes essential and must meet the requirements listed below.

Thin (several microns) cuts must be prepared using a microtome or blade (this must be done by a trained professional) along the three principal directions. Anoptical light transmission microscope (with magnification of up to 400–1500 times) is used to identify the characteristic microscopic anatomical features typical for each species or species group. This is a laboratory process requiring specialized equipment and well-trained professionals [4].

11.2 Application

Species identification is required for the estimates of mechanical (design) properties of structural members and is often combined with visual grading. Identifying the member species correctly is critical to determining accurate design properties; therefore, species identification must be always performed.

Any identification report must report the methodology used in the identification and the anatomical features utilized for the identification. The name of the identified species (or group of species) must be properly reported according to international rules that require both scientific and commercial names based on the international standards.

11.3 Limitations

Both macroscopic and microscopic species identification require significant training and experience. While it can be relatively simple to distinguish between softwood and hardwood species, in many cases it is impossible to identify the exact species by unassisted visual examination or with the aid of a magnifying glass (still regarded as a macroscopic technique). Proper species identification at a minimum requires the use of optical light microscopy.

In several cases, during the identification process, it may be impossible to identify the species and it might be necessary to terminate the identification at a higher level within the botanic taxonomy; e.g. it might be impossible to identify the individual species within the genus Populus (Poplar).

References

1. UNI 11118. Beni culturali – Manufatti lignei – Criteri per l'identificazione delle specie legnose (Cultural heritage - wooden artefacts - criteria for the identification of the wood species), 2004.
2. Kasal, B. Estimate of the design values of the in-situ wood structural members based on semi-destructive experiments. In: *Proceedings of Conservation of the Historic Wooden Structures*, Florence, Italy, February 23–27, 2005.
3. Schweingruber, F.H., *Mikroskopische holzanatomie*. EFWSL, Birmensdorf, Switzerland, 1990.
4. Nardi, B.R., La struttura anatomica del legno ed il riconoscimento dei legnami italiani di più corrente impiego, II ed., edited by S. Berti, M. Fioravanti, and N. Macchioni. CNR-IVALSA, Sesto Fiorentino, Italy, 2006.

Chapter 12
Dendrochronology

Olivia Pignatelli

Abstract Dendrochronology is a high-precision dating method for wooden elements, which applied on standing timber structures, offers a precious contribution for the reconstruction of the history of an ancient architecture and the comprehension of its evolution. The results of these investigations are of great value not only for historical studies but also for preservation, restoration and consolidation purposes of historic buildings. The paper explains the general principles and procedures, with main attention to the sampling procedures, like taking cores from standing beam or direct measurement for end sections of timbers. Furthermore, it is taken into account the accuracy of dendrochronological dating, in relation to the different type of wooden elements. Possibilities and limitations of the method are discussed.

Key words: dating, felling year, master chronology, non-destructive sampling, sapwood, tree-ring

12.1 Background

Dendrochronology is an accurate and precise method for dating wooden artifacts that in specific cases can provide the year or even the season in which the parent trees were felled. This method that was developed by the American astronomer A. Douglass at the beginning of the last century is utilized in the Cultural Heritage profession in Europe and in North America and recently in Asia.

Dendrochronological dating offers an important tool for reconstructing the history of ancient architecture and construction, and the comprehension of its evolution. The results of these investigations are of great value not only for historical studies but also for preservation, restoration and consolidation purposes. In fact, knowing the dates of the various components in a building allows investigators to assess different intervention methods according to the date of construction.

B. Kasal and T. Tannert (eds.), In Situ Assessment of Structural Timber, 109–114.

Recently a technical standard, the UNI 11141:2004, was published in Italy by the UNI-Normal Working Group 20. In this document, some guidelines for dendrochronological dating of wood in Cultural Heritage sites are proposed. To better understand the outcomes of these investigations, a brief introduction on general principles and procedures is presented, followed by a summary of different aspects of dendrochronological dating as applied to historic timber structures. More technical details of the methods and practical examples of the application of dendrochronology within Cultural Heritage programs are described in [1–8].

12.2 Equipment

Dendrochronological analyses are normally done by measuring tree-ring width on the cross-section (end grain) of a timber. For historic timber structures, generally one or two cores that can be accurately studied in the laboratory are extracted from each selected element by means of an electric or manual increment corer. The core produces a hole approximately 10 to 15 mm in diameter, depending on the specification of the corer. Areas without knots or visible defects are most suitable for coring, starting from the edge and attempting to reach the axial centre so as to obtain the longest possible ring sequences. In the rare cases in which some elements have to be replaced during repair works, slices with 5 to 10 cm thickness can be cut from the timbers.

For end sections of timbers, measurements can be made in situ by using a magnifying glass with a graduated scale after adequately preparing the surface along the selected measuring direction. If the tree-rings are clearly visible on a cross-section or radial section, other types of non-destructive sampling methods can also be carried out. For example, taking photographs or using frottage is a very simple technique consisting of transposition of the tree-ring sequence onto a paper sheet by rubbing with a pencil. These two sampling methods are indirect measurements and the actual tree-ring measurement happens later in the laboratory.

The choice of methods is a function of features of the investigated wood elements. A specific case in the examination of ancient architecture is represented by the study of wooden foundations, which can be found in towns built on water such as Venice or Amsterdam. For waterlogged wood, generally, it is necessary to cut a cross-section of the element. Depending on the wood species and the preservation of the wood, only a few cases have shown successful sampling of the cores.

During the sampling phase, particular care must be taken in the selection of the samples. Timber with regular ring patterns showing a high number of rings should be sampled because samples with only a few rings do not ensure reliable dating. In order to achieve precise results, particular care should also be devoted to identifying timbers that preserve the last ring formed under the bark or, less desirably, part of the sapwood.

Furthermore, the samples to be tested must be representative, since analyzing a single element will rarely offer significant results. Generally 10–12 samples of the same building phase should be taken [3, 5]. Analysis of several samples belonging to the same building phase allows averaging the cross-dated sequences to minimize the irregularities of the single curves thus increasing the chance to obtain a correct date.

It should also be noted that to date a building correctly, preliminary inspection of the structure should be done before sampling. This inspection should be carried out by the dendrochronologist and the architect in order to identify the original elements of the building and any reused or repaired timbers.

12.3 Application

Dendrochronology is the study and the measurement of annual tree-rings produced in regions with temperate climates. The tree-ring width is connected to various elements of the tree's genetics, the type of soil, vegetation, and other environmental factors but is most importantly linked to the climatic conditions during growth.

Dendrochronology is based on the assumption that in temperate climates, trees of the same species growing in the same geographical area during the same period of time, produce similar sequences of tree-rings, each corresponding to one year. Tree-ring widths are generally measured in a laboratory by using specialized equipment with an accuracy of 0.01 mm. The equipment consists of a stereomicroscope and travelling stage connected to a computer. Before measuring, the surface of the sample is carefully prepared in order to highlight the tree-ring's borders by using a scalpel or razor blade. The collected data are generally presented as a graph, called tree-ring curves, where the abscissa displays the chronological sequences of tree-rings and the ordinate displays the corresponding widths. Consequently, the ring sequences can be compared through an optical and statistical examination. This procedure is called cross-dating, which is created through investigation of samples from living trees whose ring dates are known up to the date of the last ring under the bark. These are compared to successively older samples, ultimately creating long term reference chronologies. These master chronologies, created with a high number of samples belonging to different time periods, show the growth pattern of certain tree species in specific geographical areas over the centuries or millennia.

In practice, an ancient wooden artifact can be dated by comparing, either visually or by using statistical tests, its tree-ring sequence against appropriate reference chronologies established for the geographical area of the wood provenance.

12.3.1 Wood Species Identification

An indispensable step in dendrochronological investigations is wood identification. This is because the various dendrochronological sequences can only be compared to timber of the same species. Furthermore, not all species are suitable for dendrochronological dating. Wood species identification should be carried out before dendrochronological sampling to define feasibility and to plan the extent of the samples to be taken. Species investigation is carried out by examining thin slices of wood taken in the three main anatomical planes (transverse, tangential and radial) under a light microscope.

12.4 Limitations

The degree of accuracy of dendrochronological dating is related to the amount of modification of the investigated wood element. In the case of circular cross-section members (where the trunk was only debarked) or partially squared members (which still have traces of bark or cambium, i.e. the last ring formed under the bark (waney edge)) it is possible to define not only the year, but also the season of the year when the element was felled. It is important to note that often the beams or the boards are stripped of the outermost rings which are removed to make the trunk square. In this case, dendrochronological dating can only identify a terminus post quem related to when the tree was felled.

In differentiated heartwood timber species, the presence of at least part of the sapwood indicates that only the outermost rings of the trunk are missing. In this case, for certain species such as oak or larch, the approximate number of missing sapwood rings can be estimated [10, 11]. Consequently, the tree-felling window and the terminus post quem of placing is narrowed. In other cases, examination of a considerable number of timbers belonging to the same building phase and accurately selected by the dendrochronologist in strict co-operation with the architect or art historian, can define a terminus near the felling date, even if the outermost portion of the trunk is missing.

Dendrochronology can determine with extreme accuracy the felling year or the season of harvest of a timber structural element, but it is important to understand that dendrochronology does not always provide accurate dating of a timber. The success of dendrochronological dating depends on: correct identification of wood species; availability of appropriate reference chronologies for the time period and geographical area; the number of tree-rings sampled, and in the case of a single sample also on the behaviour of the tree-ring sequence.

Not all wood species are appropriate for dendrochronological dating. Among the structural timbers, conifers and oak are suitable. Moreover, many master chronologies are available for these two species in different geographical regions [10, 11]. Samples with short tree-ring sequences are difficult to date. Although the lowest numbers of tree-rings to use is not fixed, in general for dating purposes it is preferable that samples contain more than 50 tree-rings [3]. The longer the tree-ring series is, the greater the chance is to obtain accurate dating. Some single samples marked by irregular growth are often not datable.

To interpret the results of dendrochronological dating, it is very important to remember that the date obtained refers to the date on which the original tree was

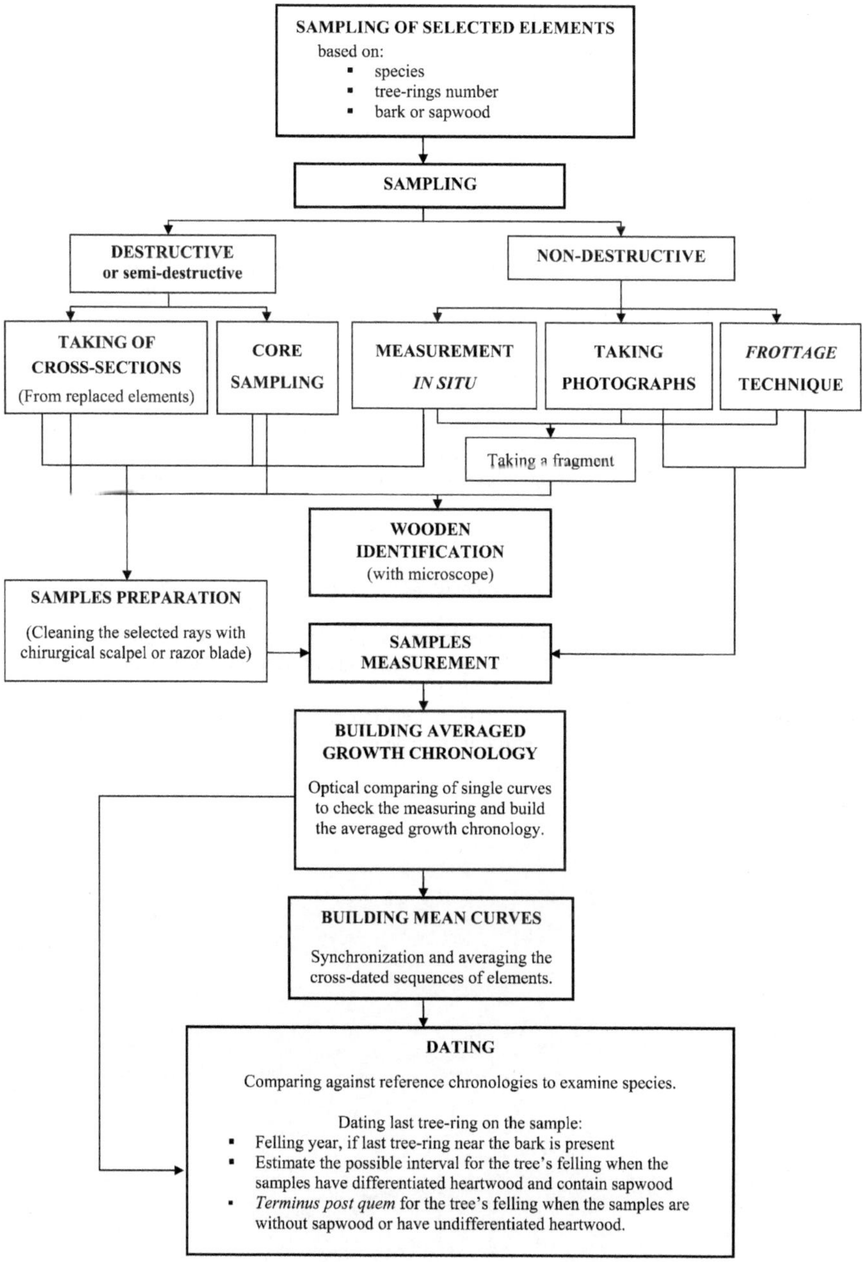

Fig. 12.1 Schematic of dendrochronological assessment

felled and not to the date of construction, even though for structural timbers there is not generally a long period of seasoning. It also should be noted that reused timbers can be presented.

References

1. Baillie, M.G.L. *Tree-Ring Dating and Archaeology*, London/Canberra, 274 pp., 1982.
2. Fritts, H.C. *Tree Rings and Climate*. Academic Press, London, 576 pp., 1976.
3. Eckstein, D., Baillie, M.G.L., and Egger, H. *European Science Foundation Handbooks for Archeologists, No. 2, Dendrochronological Dating*, 1984.
4. Hillam, J. *Dendrochronology: Guidelines on Producing and Interpreting Dendrochronological Dates*. English Heritage, London, 1998.
5. Lambert, G.N. La dendrochronologie, mémoire de l'arbre. In: *Les méthodes de datation en laboratoire*. Editions Errance, Paris, pp. 13–70, 1998.
6. Schweingruber, F.H. *Tree Rings. Basics and Applications of Dendrochronology*, Reidel, Dordrecht, 276 pp., 1988.
7. UNI 11141. Beni culturali – Manufatti lignei – Linee guida per la datazione dendrocronologica del legno, 2004.
8. Hollstein, E. *Mitteleuropaeische Eichenchronologie, Trier Grabungen und Forschungen, II*, 273 pp., 1980.
9. Corona. P. Anelli d'alburno in larice cisalpino. *Dendrochronologia*, Vol. 2, pp. 91–97, 1984.
10. Levanic, T. Update on the on-line European Catalogue of Tree-Ring Chronologies. *Dendrochronologia*, Vol. 19(2), pp. 245–251, 2001.
11. ITRDB. International Tree Ring Data Bank. Laboratory of Tree-Ring Research, University of Arizona, Tucson. URL address: http://www.ngdc.noaa.gov.

Chapter 13
Review of Codes and Standards

Nicola Macchioni, Clara Bertolini and Thomas Tannert

Abstract The chapter deals with a review of the international codes and standard related to the in situ evaluation of timber structures. The ISO 13822:2001 standard is the only one at international level on this field, but it's only partially applicable to the topic, because it deals with all the building materials, not only with wood, and because the principal aim is not the conservation criteria, but the structural safety obtained with cost saving. At national level there are two Italian standards (UNI 11119:2004 and UNI 11138:2004), both the standards are strictly related to the cultural heritage, then having as first goal the protection and conservation of the artefacts. Efforts in Switzerland aim at developing a new series of standards for the maintenance and reassessment of existing structures. A set of codes is written, including basic rules (SIA 269) and specific adaptive rules for actions on existing structures, and for building materials including timber. The main part of the chapter describes the UNI 11119:2004, defining the mechanical performances of each structural member at the moment of the diagnostic inspection. The steps of the diagnostic path start from the identification of the species, pass through the description of the existing conditions with an accurate constructive survey and the inspection procedures, going finally to the inspection report that describes the results.

Key words: codes, dendrochronology, diagnosis, grading, inspection, report, standard, survey, timber structures

13.1 Codes and Standards

13.1.1 International Standards

Despite relatively large numbers of technical papers, few international standards related to in situ evaluation of timber exist. The chapter deals with a review of the

B. Kasal and T. Tannert (eds.), In Situ Assessment of Structural Timber, 115–121.

international codes and standard related to the in situ evaluation of timber structures. The ISO 13822:2001 standard is the only one at international level on this field, but it is only partially applicable to the topic, because it deals with all the building materials, not only with wood, and because the principal aim is not the conservation criteria, but the structural safety obtained with cost saving. At national level there are two Italian standards (UNI 11119:2004 and UNI 11138:2004), both the standards are strictly related to the cultural heritage, then having as first goal the protection and conservation of the artefacts. Efforts in Switzerland aim at developing a new series of standards for the maintenance and reassessment of existing structures. A set of codes is written, including basic rules (SIA 269) and specific adaptive rules for actions on existing structures, and for building materials including timber. The main part of the chapter describes the UNI 11119:2004, defining the mechanical performances of each structural member at the moment of the diagnostic inspection. The steps of the diagnostic path start from the identification of the species, pass through the description of the existing conditions with an accurate constructive survey and the inspection procedures, going finally to the inspection report that describes the results. At an international level the standard ISO 13822 [2] "Bases for Design of Structures – Assessment of Existing Structures" is the closest to this topic. The introduction states:

> This document is intended not only as a standard of principles and procedures for the assessment of existing structures but also as a guide for use by structural engineers and clients. Engineers can apply specific methods for assessment to save structures and to reduce a client's expenditure. The goal is to limit construction intervention to a strict minimum, a goal that is clearly in agreement with the principles of sustainable development.

Clearly, the adaptability of this standard to the in situ evaluation of timber members is only partial. The standard is intended for all types of existing structures and materials (steel, concrete, masonry and timber). The standard focus is safety, economy and in situ evaluation of historic timber structures and focuses on conservation of cultural heritage:

> This International Standard is applicable to historical structures, provided additional considerations are taken into account concerning the preservation of the historical appearance of the structure and the preservation of its historical materials.

CEN, the European standardization body, started the activity on the cultural heritage topic at the end of 2004 (TC 346 – Conservation of cultural properties), but no standards have been produced and no research on in situ evaluation of timber elements has been published.

13.1.2 National Standards

At a national level, Italy has a complete series of standards regarding the approach to the wooden artifacts, including timber structures, belonging to cultural heritage sites. UNI 11161:2005 [3] establishes requirements for conservation, maintenance and restoration of wooden artifacts that are a part of Italy's cultural heritage program. This standard defines essential criteria that must be followed when carrying out interventions. It is applicable for all types of wooden artifacts including timber structures.

The following items are covered by the document mentioned above: historical documentation; chronological documentation of interventions and changes; object description and photographic documentation; graphical representation; object dating; description of used methodology; wood species identification; environmental thermo-hygrometric conditions and moisture content of wood documentation; description of the environmental thermo-hygrometric conditions after intervention; condition diagnosis of all wood components, and classification and quantification of potential decay.

These eight points can be separated into two groups: (1) the description of the present state (historical documentation, description, survey and graphical representation of the object, and dating), and (2) diagnosis of the elements (wood identification, dating, existing and future environmental conditions).

In the US, no standards specific to in situ evaluation of timber exist. The ASCE Standard [4] covers the most fundamental aspects of evaluation of existing buildings but does not offer enough information to be of any practical significance. Material-specific codes are used to establish, for example, moisture contents, density or some mechanical properties but these codes are only applicable if prescribed specimens can be extracted from the structure. Efforts in Switzerland, that started in 2006 and will conclude in 2011, develop a new series of standards for the maintenance and reassessment of existing structures. The basic rules for the assessment and maintenance of existing structures form the Draft Swiss Code SIA 269 [5] that includes the following parts: (i) General (examination, monitoring and maintenance, economic and cultural value); (ii) Requirements (use, structural safety, serviceability, effectiveness of maintenance interventions); (iii) Updating (actions, properties and condition of construction products and soil/foundation, structural model, geometry, ultimate resistance, deformations); (iv) Structural analysis and verification (deterministic and probabilistic); (v) Examination (procedure, condition survey and evaluation, recommendation of interventions); (vi) Maintenance interventions (concept, realisation, monitoring, maintenance, immediate measures, additional measures regarding safety, repair, modification); and (vii) Construction documents (service criteria agreement, service instructions, basis of design, history of the structure, hazard events, monitoring and maintenance plan, inspection reports, result of monitoring, documents resulting from examination and maintenance interventions, record/plans of construction). Furthermore, specific adaptive rules for actions on existing structures and for building materials, including timber [6] are being developed.

13.2 Description of the Existing Conditions

No standards are currently available on this subject. The technological-structural survey must identify and understand the reference system through the complete analysis of the building's characteristics and structural subsystems. This activity represents a "redesign" of the structure with the goal of understanding the system from structural point of view. Such understanding is essential for potential future remediation work and/or alterations if necessary or permitted. The survey is a complex information system that includes observations, understanding of the structure, measurements, interpretation and, graphic representation of the object. Geometry and morphology of the investigated structure represent the initial data set that is used as a framework for subsequent data organization.

The initial survey must include:

- photographic survey of all the elements and all the structural connections;
- identification of structural typologies;
- graphic representation of the building details drawn to a scale;
- dimensions of all elements and relationships between them;
- representation of the data as an abacus;
- overlapping on the abaci of the obtained data with the results of the in-situ inspections and drawings of the thematic and synthesis charts.

Properly organized and performed survey and graphic representation are necessary for a successful on-site diagnosis and design of an eventual restoration plan. The survey described above provides fundamental information about the investigated system that must be complemented by complete identification of details and connections. The typologies of connections (such as mortise and tenon, dovetail, half wood, mechanical fasteners, etc.) and relevant dimensions must be documented.

The structural geometry survey is then complemented by results of the physical and biological survey that includes species identification, defects (nature, position and dimensions), biotic attack (nature, extension and position), and structural damages.

13.2.1 Diagnosis of the Structural Elements

Wood identification is discussed in Chapter 12, including specifics listed in [7] for the Italian national standard. One of the most precise dating techniques is dendrochronology. A limitation of dendrochronological dating is that it dates the material and not the structure. Italian standards [8] describe the methodology that applies to dendrochronological dating. The diagnosis is described in [8] with further discussion in [9]. The following represents the main parts of the referenced standard [10].

The goal of this standard is to establish objectives, procedures and criteria for inspection that will be used to evaluate the state of conservation and the performances of timber members in load-bearing historic timber structures.

13.2.2 Objectives of the Inspection

The inspection must evaluate the original characteristics of each timber member and modifications that each member underwent during its service life. The collected information must include:

- wood species;
- wood moisture contents and moisture gradients;
- class of biological risk, according to EN 335-1 [11] and EN 335-2 [12];
- geometry and morphology timber members indicating position and extension of defects, decay or possible damage;
- position, shape and dimension of critical area and critical cross-section;
- strength grading of the wood member and/or member in critical areas.

To conduct the inspection correctly, the object must be accessible, clean and visible.

13.2.3 Inspection Procedure

The procedure involves the species identification, non-destructive moisture content measurements via, and determination of biological risk according to standards [11, 12].

The geometric survey must include structural system and individual member characteristics (such as position of the pith, growth irregularities, type, length and position of defects), and any other information necessary for estimating the mechanical characteristics of investigated individual elements.

In order to evaluate the mechanical performance of each timber member a "critical area" must be identified. The critical area is defined as

> part of a wood element with longitudinal axes no less than 150 mm, which is considered to be relevant because of defects, position, state of conservation and stress conditions that are determined by static analysis.

Critical areas are then used to apply grading rules and determine the grade of the structural member.

Visual grading can only be used if the investigate element (such as a beam or column) is visible. If there is no access to a structural member, other investigative methods such as those described in this report must be used. The methods must be non-invasive so that structural and non-structural characteristics are not compromised.

The following general criteria need to be applied during grading:

- grade the entire wood member and if necessary, identify each critical area separately;
- account for potential lack of full access impairing the ability of inspecting the entire member surface area;
- if an alteration occurs due to mechanical damage or localized biological decay (rot, insect attacks found on the surface) use the efficient section only.

13.2.4 Inspection Report

The inspection report must include:

- a description of the structure;
- specific objectives of the inspection;
- a period in which the inspection was carried out (date);
- a description of instrumentation used;
- results;
- the name, qualification and signature of the person responsible for the inspection.

The discussed standard [11] also includes a table that gives the maximum values of stresses that can be used applying the allowable stress design methods and average modulus of elasticity (E) for each category and wood species.

In situ diagnosis is also discussed in [13]. This standard specifically describes the need of a preventative evaluation of the present state of the artifact in order to understand the performance of the overall static suitability of a building and the role of the timber structure within the building. Keeping this in mind, the sole purpose of in situ diagnosis is to provide information and unforeseen results do not necessarily mean that the execution of a restoration intervention must be carried out.

Historical analysis is a very important aspect of assessment that reveals historical events affecting a structure such as its structural typology and evolution, construction characteristics and traumatic events. A dendrochronological analysis of wood members can play an important role in the historical analysis.

References

1. Diamantidis, D. *Probabilistic Assessment of Existing Structures*, The Joint Committee on Structural Safety (JCSS). RILEM Publications, 2001.
2. ISO 13822:2001(E). Bases for design of structures – Assessment of existing structures. International Standard Organisation, Geneva (CH), 2001.
3. UNI 11161:2005. Beni culturali – Manufatti lignei – Linee guida per la conservazione, la manutenzione e il restauro. UNI, Milano (I), 2005.
4. ASCE 11-99. Guideline for Structural Condition Assessment of Existing Buildings, SEI/ASCE 11-99. American Society of Civil Engineers, 160 pp., 1999.

5. Normentwurf SIA 269:2009, Grundlagen der Erhaltung von Tragwerken (Draft Standard SIA 269:2009 Basis of Maintenance of Structures), Swiss Society of Engineers and Architects SIA, Zurich, Switzerland, 2009.
6. Normentwurf SIA 269/5:2009 Erhaltung von Tragwerken – Holzbau (Draft Standard SIA 269/5:2009 Existing Structures – Timber Structures), Swiss Society of Engineers and Architects SIA, Zurich, Switzerland, 2009.
7. UNI 11118:2004. Cultural heritage – wooden artefacts – Criteria for the identification of wooden species/Beni culturali – Manufatti lignei – Criteri per l'identificazione delle specie legnose. UNI, Milano (I), 2004.
8. UNI 11141:2004. Beni culturali – Manufatti lignei – Linee guida per la datazione dendrocronologica del legno. UNI, Milano (I), 2004.
9. Macchioni, N. and Piazza, M. Italian standardisation activity in the field of diagnosis and restoration of ancient timber structures, In: *Structural Analysis of Historical Constructions*, P.B. Lourenço, P. Roca, C. Modena, S. Agrawal (Eds.), New Delhi, 2006.
10. UNI 11119:2004. Cultural heritage – Wooden artefacts – Load bearing structures of buildings – On site inspection for the diagnosis of timber members (Beni culturali – Manufatti lignei – Strutture portanti degli edifici – Ispezioni in-situ per la diagnosi degli elementi in opera). UNI, Milano (I), 2004.
11. EN 335-1. Durability of wood and wood-based products – Definition of use classes – Part 1: General. CEN, Brussels (B), 2006.
12. EN 335-2. Durability of wood and wood-based products – Definition of use classes – Part 2: Application to solid wood. CEN, Brussels (B), 2006.
13. UNI 11138. Cultural heritage – Load bearing structures of buildings – Criteria for the preliminary evaluation, design and execution of works (Beni culturali – Manufatti lignei – Strutture portanti degli edifici – Criteri per la valutazione preventiva, la progettazione e l'esecuzione di interventi). UNI, Milano (I), 2004.

Key Words Index

REUNION INTERNATIONALE DES LABORATOIRES ET EXPERTS
DES MATERIAUX, SYSTEMES DE CONSTRUCTION ET OUVRAGES

INTERNATIONAL UNION OF LABORATORIES AND EXPERTS IN CONSTRUCTION MATERIALS, SYSTEMS AND STRUCTURES

RILEM Publications – 27 July 2010

The following list is presenting our global offer, sorted by series.

RILEM PROCEEDINGS

PRO 1: Durability of High Performance Concrete (ISBN: 2-912143-03-9); *Ed. H. Sommer*

PRO 2: Chloride Penetration into Concrete (ISBN: 2-912143-00-04); *Eds. L.-O. Nilsson and J.-P. Ollivier*

PRO 3: Evaluation and Strengthening of Existing Masonry Structures (ISBN: 2-912143-02-0); *Eds. L. Binda and C. Modena*

PRO 4: Concrete: From Material to Structure (ISBN: 2-912143-04-7); *Eds. J.-P. Bournazel and Y. Malier*

PRO 5: The Role of Admixtures in High Performance Concrete (ISBN: 2-912143-05-5); *Eds. J. G. Cabrera and R. Rivera-Villarreal*

PRO 6: High Performance Fiber Reinforced Cement Composites - HPFRCC 3 (ISBN: 2-912143-06-3); *Eds. H. W. Reinhardt and A. E. Naaman*

PRO 7: 1st International RILEM Symposium on Self-Compacting Concrete (ISBN: 2-912143-09-8); *Eds. Å. Skarendahl and Ö. Petersson*

PRO 8: International RILEM Symposium on Timber Engineering (ISBN: 2-912143-10-1); *Ed. L. Boström*

PRO 9: 2nd International RILEM Symposium on Adhesion between Polymers and Concrete ISAP '99 (ISBN: 2-912143-11-X); *Eds. Y. Ohama and M. Puterman*

PRO 10: 3rd International RILEM Symposium on Durability of Building and Construction Sealants (ISBN: 2-912143-13-6); *Eds. A. T. Wolf*

PRO 11: 4th International RILEM Conference on Reflective Cracking in Pavements (ISBN: 2-912143-14-4); *Eds. A. O. Abd El Halim, D. A. Taylor and El H. H. Mohamed*

PRO 12: International RILEM Workshop on Historic Mortars: Characteristics and Tests (ISBN: 2-912143-15-2); *Eds. P. Bartos, C. Groot and J. J. Hughes*

PRO 13: 2nd International RILEM Symposium on Hydration and Setting (ISBN: 2-912143-16-0); *Ed. A. Nonat*

PRO 14: Integrated Life-Cycle Design of Materials and Structures - ILCDES 2000 (ISBN: 951-758-408-3); (ISSN: 0356-9403); *Ed. S. Sarja*

PRO 15: Fifth RILEM Symposium on Fibre-Reinforced Concretes (FRC) - BEFIB'2000 (ISBN: 2-912143-18-7); *Eds. P. Rossi and G. Chanvillard*

PRO 16: Life Prediction and Management of Concrete Structures (ISBN: 2-912143-19-5); *Ed. D. Naus*

PRO 17: Shrinkage of Concrete – Shrinkage 2000 (ISBN: 2-912143-20-9); *Eds. V. Baroghel-Bouny and P.-C. Aïtcin*

PRO 18: Measurement and Interpretation of the On-Site Corrosion Rate (ISBN: 2-912143-21-7); *Eds. C. Andrade, C. Alonso, J. Fullea, J. Polimon and J. Rodriguez*

PRO 19: Testing and Modelling the Chloride Ingress into Concrete (ISBN: 2-912143-22-5); *Eds. C. Andrade and J. Kropp*

REUNION INTERNATIONALE DES LABORATOIRES ET EXPERTS
DES MATERIAUX, SYSTEMES DE CONSTRUCTION ET OUVRAGES

INTERNATIONAL UNION OF LABORATORIES AND EXPERTS IN CONSTRUCTION MATERIALS, SYSTEMS AND STRUCTURES

PRO 20: 1st International RILEM Workshop on Microbial Impacts on Building Materials (CD 02) (e-ISBN 978-2-35158-013-4); *Ed. M. Ribas Silva*

PRO 21: International RILEM Symposium on Connections between Steel and Concrete (ISBN: 2-912143-25-X); *Ed. R. Eligehausen*

PRO 22: International RILEM Symposium on Joints in Timber Structures (ISBN: 2-912143-28-4); *Eds. S. Aicher and H.-W. Reinhardt*

PRO 23: International RILEM Conference on Early Age Cracking in Cementitious Systems (ISBN: 2-912143-29-2); *Eds. K. Kovler and A. Bentur*

PRO 24: 2nd International RILEM Workshop on Frost Resistance of Concrete (ISBN: 2-912143-30-6); *Eds. M. J. Setzer, R. Auberg and H.-J. Keck*

PRO 25: International RILEM Workshop on Frost Damage in Concrete (ISBN: 2-912143-31-4); *Eds. D. J. Janssen, M. J. Setzer and M. B. Snyder*

PRO 26: International RILEM Workshop on On-Site Control and Evaluation of Masonry Structures (ISBN: 2-912143-34-9); *Eds. L. Binda and R. C. de Vekey*

PRO 27: International RILEM Symposium on Building Joint Sealants (CD03); *Ed. A. T. Wolf*

PRO 28: 6th International RILEM Symposium on Performance Testing and Evaluation of Bituminous Materials - PTEBM'03 (ISBN: 2-912143-35-7; e-ISBN: 978-2-912143-77-8); *Ed. M. N. Partl*

PRO 29: 2nd International RILEM Workshop on Life Prediction and Ageing Management of Concrete Structures (ISBN: 2-912143-36-5); *Ed. D. J. Naus*

PRO 30: 4th International RILEM Workshop on High Performance Fiber Reinforced Cement Composites - HPFRCC 4 (ISBN: 2-912143-37-3); *Eds. A. E. Naaman and H. W. Reinhardt*

PRO 31: International RILEM Workshop on Test and Design Methods for Steel Fibre Reinforced Concrete: Background and Experiences (ISBN: 2-912143-38-1); *Eds. B. Schnütgen and L. Vandewalle*

PRO 32: International Conference on Advances in Concrete and Structures 2 vol. (ISBN (set): 2-912143-41-1); *Eds. Ying-shu Yuan, Surendra P. Shah and Heng-lin Lü*

PRO 33: 3rd International Symposium on Self-Compacting Concrete (ISBN: 2-912143-42-X); *Eds. Ó. Wallevik and I. Níelsson*

PRO 34: International RILEM Conference on Microbial Impact on Building Materials (ISBN: 2-912143-43-8); *Ed. M. Ribas Silva*

PRO 35: International RILEM TC 186-ISA on Internal Sulfate Attack and Delayed Ettringite Formation (ISBN: 2-912143-44-6); *Eds. K. Scrivener and J. Skalny*

PRO 36: International RILEM Symposium on Concrete Science and Engineering – A Tribute to Arnon Bentur (ISBN: 2-912143-46-2); *Eds. K. Kovler, J. Marchand, S. Mindess and J. Weiss*

PRO 37: 5th International RILEM Conference on Cracking in Pavements – Mitigation, Risk Assessment and Prevention (ISBN: 2-912143-47-0); *Eds. C. Petit, I. Al-Qadi and A. Millien*

PRO 38: 3rd International RILEM Workshop on Testing and Modelling the Chloride Ingress into Concrete (ISBN: 2-912143-48-9); *Eds. C. Andrade and J. Kropp*

PRO 39: 6th International RILEM Symposium on Fibre-Reinforced Concretes - BEFIB 2004 (ISBN: 2-912143-51-9); *Eds. M. Di Prisco, R. Felicetti and G. A. Plizzari*

PRO 40: International RILEM Conference on the Use of Recycled Materials in Buildings and Structures (ISBN: 2-912143-52-7); *Eds. E. Vázquez, Ch. F. Hendriks and G. M. T. Janssen*

PRO 41: RILEM International Symposium on Environment-Conscious Materials and Systems for Sustainable Development (ISBN: 2-912143-55-1); *Eds. N. Kashino and Y. Ohama*

REUNION INTERNATIONALE DES LABORATOIRES ET EXPERTS
DES MATERIAUX, SYSTEMES DE CONSTRUCTION ET OUVRAGES

INTERNATIONAL UNION OF LABORATORIES AND EXPERTS IN CONSTRUCTION MATERIALS, SYSTEMS AND STRUCTURES

PRO 42: SCC'2005 - China: 1st International Symposium on Design, Performance and Use of Self-Consolidating Concrete (ISBN: 2-912143-61-6); *Eds. Zhiwu Yu, Caijun Shi, Kamal Henri Khayat and Youjun Xie*

PRO 43: International RILEM Workshop on Bonded Concrete Overlays (e-ISBN: 2-912143-83-7); *Eds. J. L. Granju and J. Silfwerbrand*

PRO 44: 2nd International RILEM Workshop on Microbial Impacts on Building Materials (CD11) (e-ISBN: 2-912143-84-5); *Ed. M. Ribas Silva*

PRO 45: 2nd International Symposium on Nanotechnology in Construction, Bilbao (ISBN: 2-912143-87-X); *Eds. Peter J. M. Bartos, Yolanda de Miguel and Antonio Porro*

PRO 46: ConcreteLife'06 - International RILEM-JCI Seminar on Concrete Durability and Service Life Planning: Curing, Crack Control, Performance in Harsh Environments (ISBN: 2-912143-89-6); *Ed. K. Kovler*

PRO 47: International RILEM Workshop on Performance Based Evaluation and Indicators for Concrete Durability (ISBN: 978-2-912143-95-2); *Eds. V. Baroghel-Bouny, C. Andrade, R. Torrent and K. Scrivener*

PRO 48: 1st International RILEM Symposium on Advances in Concrete through Science and Engineering (e-ISBN: 2-912143-92-6); *Eds. J. Weiss, K. Kovler, J. Marchand, and S. Mindess*

PRO 49: International RILEM Workshop on High Performance Fiber Reinforced Cementitious Composites in Structural Applications (ISBN: 2-912143-93-4); *Eds. G. Fischer and V.C. Li*

PRO 50: 1st International RILEM Symposium on Textile Reinforced Concrete (ISBN: 2-912143-97-7); *Eds. Josef Hegger, Wolfgang Brameshuber and Norbert Will*

PRO 51: 2nd International Symposium on Advances in Concrete through Science and Engineering (ISBN: 2-35158-003-6; e-ISBN: 2-35158-002-8); *Eds. J. Marchand, B. Bissonnette, R. Gagné, M. Jolin and F. Paradis*

PRO 52: Volume Changes of Hardening Concrete: Testing and Mitigation (ISBN: 2-35158-004-4; e-ISBN: 2-35158-005-2); *Eds. O. M. Jensen, P. Lura and K. Kovler*

PRO 53: High Performance Fiber Reinforced Cement Composites - HPFRCC5 (ISBN: 978-2-35158-046-2); *Eds. H. W. Reinhardt and A. E. Naaman*

PRO 54: 5th International RILEM Symposium on Self-Compacting Concrete (ISBN: 978-2-35158-047-9); *Eds. G. De Schutter and V. Boel*

PRO 55: International RILEM Symposium Photocatalysis, Environment and Construction Materials (ISBN: 978-2-35158-056-1); *Eds. P. Baglioni and L. Cassar*

PRO56: International RILEM Workshop on Integral Service Life Modelling of Concrete Structures (ISBN 978-2-35158-058-5); *Eds. R. M. Ferreira, J. Gulikers and C. Andrade*

PRO57: RILEM Workshop on Performance of cement-based materials in aggressive aqueous environments (e-ISBN: 978-2-35158-059-2); *Ed. N. De Belie*

PRO58: International RILEM Symposium on Concrete Modelling - CONMOD'08 (ISBN: 978-2-35158-060-8); *Eds. E. Schlangen and G. De Schutter*

PRO 59: International RILEM Conference on On Site Assessment of Concrete, Masonry and Timber Structures - SACoMaTiS 2008 (ISBN set: 978-2-35158-061-5); *Eds. L. Binda, M. di Prisco and R. Felicetti*

PRO 60: Seventh RILEM International Symposium on Fibre Reinforced Concrete: Design and Applications - BEFIB 2008 (ISBN: 978-2-35158-064-6); *Ed. R. Gettu*

REUNION INTERNATIONALE DES LABORATOIRES ET EXPERTS
DES MATERIAUX, SYSTEMES DE CONSTRUCTION ET OUVRAGES

INTERNATIONAL UNION OF LABORATORIES AND EXPERTS IN CONSTRUCTION MATERIALS, SYSTEMS AND STRUCTURES

PRO 61: 1st International Conference on Microstructure Related Durability of Cementitious Composites 2 vol., (ISBN: 978-2-35158-065-3); *Eds. W. Sun, K. van Breugel, C. Miao, G. Ye and H. Chen*

PRO 62: NSF/ RILEM Workshop: In-situ Evaluation of Historic Wood and Masonry Structures (e-ISBN: 978-2-35158-068-4); *Eds. B. Kasal, R. Anthony and M. Drdácký*

PRO 63: Concrete in Aggressive Aqueous Environments: Performance, Testing and Modelling, 2 vol., (ISBN: 978-2-35158-071-4); *Eds. M. G. Alexander and A. Bertron*

PRO 64: Long Term Performance of Cementitious Barriers and Reinforced Concrete in Nuclear Power Plants and Waste Management - NUCPERF 2009 (ISBN: 978-2-35158-072-1); *Eds. V. L'Hostis, R. Gens, C. Gallé*

PRO 65: Design Performance and Use of Self-consolidating Concrete - SCC'2009 (ISBN: 978-2-35158-073-8); *Eds. C. Shi, Z. Yu, K. H. Khayat and P. Yan*

PRO 66: 2nd International RILEM Workshop on Concrete Durability and Service Life Planning - ConcreteLife'09 (ISBN: 978-2-35158-074-5); *Ed. K. Kovler*

PRO 67: Repairs Mortars for Historic Masonry (e-ISBN: 978-2-35158-083-7); *Ed. C. Groot*

PRO 68: Proceedings of the 3rd International RILEM Symposium on 'Rheology of Cement Suspensions such as Fresh Concrete (ISBN 978-2-35158-091-2); *Eds. O. H. Wallevik, S. Kubens and S. Oesterheld*

PRO 69: 3rd International PhD Student Workshop on 'Modelling the Durability of Reinforced Concrete (ISBN: 978-2-35158-095-0); *Eds. R. M. Ferreira, J. Gulikers and C. Andrade*

PRO 71: Advances in Civil Engineering Materials - The 50-year Teaching Anniversary of Prof. Sun Wei' (ISBN: 978-2-35158-098-1; e-ISBN: 978-2-35158-099-8); *Eds. C. Miao, G. Ye, and H. Chen*

RILEM REPORTS

Report 19: Considerations for Use in Managing the Aging of Nuclear Power Plant Concrete Structures (ISBN: 2-912143-07-1); *Ed. D. J. Naus*

Report 20: Engineering and Transport Properties of the Interfacial Transition Zone in Cementitious Composites (ISBN: 2-912143-08-X); *Eds. M. G. Alexander, G. Arliguie, G. Ballivy, A. Bentur and J. Marchand*

Report 21: Durability of Building Sealants (ISBN: 2-912143-12-8); *Ed. A. T. Wolf*

Report 22: Sustainable Raw Materials - Construction and Demolition Waste (ISBN: 2-912143-17-9); *Eds. C. F. Hendriks and H. S. Pietersen*

Report 23: Self-Compacting Concrete state-of-the-art report (ISBN: 2-912143-23-3); *Eds. Å. Skarendahl and Ö. Petersson*

Report 24: Workability and Rheology of Fresh Concrete: Compendium of Tests (ISBN: 2-912143-32-2); *Eds. P. J. M. Bartos, M. Sonebi and A. K. Tamimi*

Report 25: Early Age Cracking in Cementitious Systems (ISBN: 2-912143-33-0); *Ed. A. Bentur*

Report 26: Towards Sustainable Roofing (Joint Committee CIB/RILEM) (CD 07) (e-ISBN 978-2-912143-65-5); *Eds. Thomas W. Hutchinson and Keith Roberts*

Report 27: Condition Assessment of Roofs (Joint Committee CIB/RILEM) (CD 08) (e-ISBN 978-2-912143-66-2); *Ed. CIB W 83/RILEM TC166-RMS*

Report 28: Final report of RILEM TC 167-COM 'Characterisation of Old Mortars with Respect to Their Repair (ISBN: 978-2-912143-56-3); *Eds. C. Groot, G. Ashall and J. Hughes*

REUNION INTERNATIONALE DES LABORATOIRES ET EXPERTS
DES MATERIAUX, SYSTEMES DE CONSTRUCTION ET OUVRAGES

INTERNATIONAL UNION OF LABORATORIES AND EXPERTS IN CONSTRUCTION MATERIALS, SYSTEMS AND STRUCTURES

Report 29: Pavement Performance Prediction and Evaluation (PPPE): Interlaboratory Tests (e-ISBN: 2-912143-68-3); *Eds. M. Partl and H. Piber*

Report 30: Final Report of RILEM TC 198-URM 'Use of Recycled Materials' (ISBN: 2-912143-82-9; e-ISBN: 2-912143-69-1); *Eds. Ch. F. Hendriks, G. M. T. Janssen and E. Vázquez*

Report 31: Final Report of RILEM TC 185-ATC 'Advanced testing of cement-based materials during setting and hardening' (ISBN: 2-912143-81-0; e-ISBN: 2-912143-70-5); *Eds. H. W. Reinhardt and C. U. Grosse*

Report 32: Probabilistic Assessment of Existing Structures. A JCSS publication (ISBN 2-912143-24-1); *Ed. D. Diamantidis*

Report 33: State-of-the-Art Report of RILEM Technical Committee TC 184-IFE 'Industrial Floors' (ISBN 2-35158-006-0); *Ed. P. Seidler*

Report 34: Report of RILEM Technical Committee TC 147-FMB 'Fracture mechanics applications to anchorage and bond' Tension of Reinforced Concrete Prisms – Round Robin Analysis and Tests on Bond (e-ISBN 2-912143-91-8); *Eds. L. Elfgren and K. Noghabai*

Report 35: Final Report of RILEM Technical Committee TC 188-CSC 'Casting of Self Compacting Concrete' (ISBN 2-35158-001-X; e-ISBN: 2-912143-98-5); *Eds. Å. Skarendahl and P. Billberg*

Report 36: State-of-the-Art Report of RILEM Technical Committee TC 201-TRC 'Textile Reinforced Concrete' (ISBN 2-912143-99-3); *Ed. W. Brameshuber*

Report 37: State-of-the-Art Report of RILEM Technical Committee TC 192-ECM 'Environment-conscious construction materials and systems' (ISBN: 978-2-35158-053-0); *Eds. N. Kashino, D. Van Gemert and K. Imamoto*

Report 38: State-of-the-Art Report of RILEM Technical Committee TC 205-DSC 'Durability of Self-Compacting Concrete' (ISBN: 978-2-35158-048-6); *Eds. G. De Schutter and K. Audenaert*

Report 39: Final Report of RILEM Technical Committee TC 187-SOC 'Experimental determination of the stress-crack opening curve for concrete in tension' (ISBN 978-2-35158-049-3); *Ed. J. Planas*

Report 40: State-of-the-Art Report of RILEM Technical Committee TC 189-NEC 'Non-Destructive Evaluation of the Penetrability and Thickness of the Concrete Cover' (ISBN 978-2-35158-054-7); *Eds. R. Torrent and L. Fernández Luco*

Report 41: State-of-the-Art Report of RILEM Technical Committee TC 196-ICC 'Internal Curing of Concrete' (ISBN 978-2-35158-009-7); *Eds. K. Kovler and O. M. Jensen*

Report 42: 'Acoustic Emission and Related Non-destructive Evaluation Techniques for Crack Detection and Damage Evaluation in Concrete' - Final Report of RILEM Technical Committee 212-ACD (e-ISBN: 978-2-35158-100-1); *Ed. M. Ohtsu*

RILEM COMPENDIUMS

COMP 01: Trilingual Dictionary for Materials and Structures (English-French-German) (CD01) (1970)

COMP 02: 1947-1997: 50 years of evolution of Building Materials and Structures (e-ISBN: 2-912143-86-1); *Ed. F. Wittmann*

COMP 03: General Conference of RILEM TCs' Chairmen and RILEM Seminar 'Advancing the Knowledge in Materials and Structures' (CD10) (e-ISBN: 2-912143-85-3)

COMP 06: Concrete Science and Engineering Journal – Vols. 1, 2, 3, 4 (1999-2002) (CD05)

COMP 13: RILEM Technical Day, Moscow (e-ISBN: 2-35158-045-1)

GPSR Compliance

The European Union's (EU) General Product Safety Regulation (GPSR) is a set of rules that requires consumer products to be safe and our obligations to ensure this.

If you have any concerns about our products, you can contact us on ProductSafety@springernature.com

In case Publisher is established outside the EU, the EU authorized representative is:

Springer Nature Customer Service Center GmbH
Europaplatz 3
69115 Heidelberg, Germany

Batch number: 10584448

Printed by Printforce, the Netherlands